Imagining Space

"Development of the space station is as inevitable as the rising of the sun; man has already poked his nose into space and he is not likely to pull it back. . . .

There can be no thought of finishing, for aiming at the stars—both literally and figuratively—is the work of generations, and no matter how much progress one makes, there is always the thrill of just beginning."

—Wernher von Braun, rocket scientist, 1952.

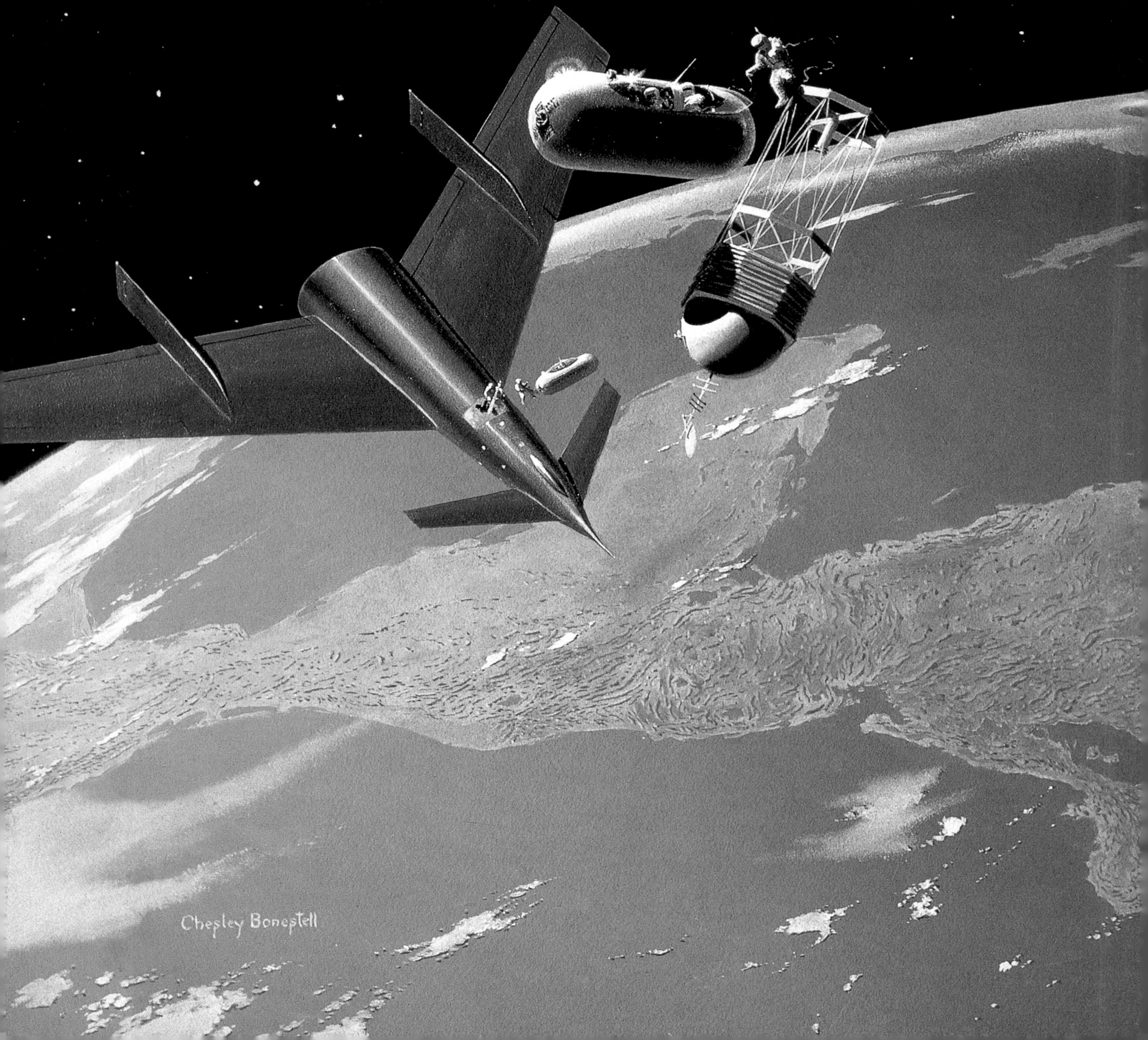

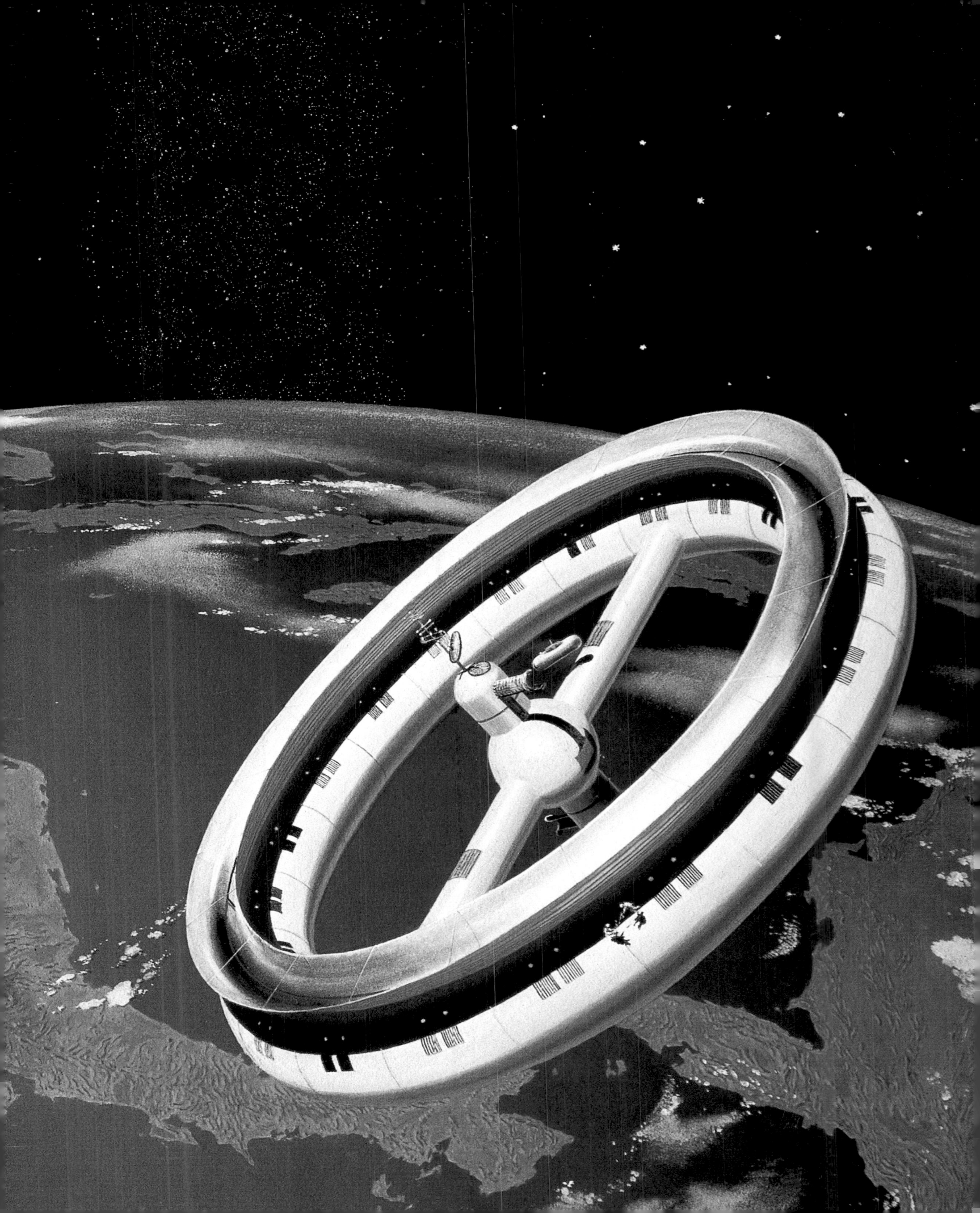

"We must somehow keep the dreams of space exploration alive, for in the long run they will prove to be of far more importance to the human race than the attainment of immediate material benefits. Like Darwin, we have set sail upon an ocean: the cosmic sea of the Universe. There can be no turning back. To do so could well prove to be a guarantee of extinction. When a nation, or a race or a planet turns its back on the future, to concentrate on the present, it cannot see what lies ahead. It can neither plan nor prepare for the future, and thus discards the vital opportunity for determining its evolutionary heritage and perhaps its survival."

—James C. Fletcher, NASA administrator, 1975.

"I want to start some businesses on the Moon, near-Earth Asteroids, and Mars. There is a lot of important science to be done on the surface of those bodies, and scientific information is valuable because it is so expensive to collect. This information has no mass but has a very high value, and transmitting it back to Earth is easy and inexpensive. We can make a profit by supplying more science per dollar than feasible today, and that is important, because if humanity wants to go to space to stay, space has to pay."

—James William Benson, chairman, SpaceDev, Inc., 1997.

NASA

"What a splendid perspective contact with a profoundly different civilization might provide! In a cosmic setting vast and old beyond ordinary human understanding we are a little lonely, and we ponder the ultimate significance, if any, of our tiny but exquisite blue planet, the Earth. . . . In the deepest sense the search for extraterrestrial intelligence is a search for ourselves."

—Carl Sagan, astrophysicist and astronomer, 1979.

Imagining Space

Achievements ★ Predictions ★ Possibilities
1950–2050

Roger D. Launius and
Howard E. McCurdy

Foreword by Ray Bradbury

CHRONICLE BOOKS
SAN FRANCISCO

DEDICATION

To Kevin Howard McCurdy, Jeremy and Justin Lenz, Nathan Patrick Winter, Kara and Blake Launius, and Hayley and Chad Yates . . . and all the other children of the new generation who will experience wonders in space yet undreamed of.

Library of Congress Cataloging-in-Publication Data available.

ISBN: 0-8118-3115-9

Printed in China

Designed by Laura Lovett
Typeset in Baskerville Book, Profile Sans, and Officina Serif

Distributed in Canada by Raincoast Books
9050 Shaughnessy Street
Vancouver, British Columbia V6P 6E5

10 9 8 7 6 5 4 3 2 1

Chronicle Books LLC
85 Second Street Street
San Francisco, California 94105

www.chroniclebooks.com

p. 1. Model of an orbital spaceplane, 1954. For decades, humans have dreamed of constructing winged spaceships capable of carrying people into orbit with as much ease as airplanes fly them through the atmosphere.

pp. 2–3. Artist's depiction of a rotating space station, 1952. Chesley Bonestell painted this famous space station panorama for the March 22, 1952, *Collier's* magazine. A winged space shuttle and a space observatory orbit nearby.

pp. 4–5. Pillars of creation, photograph taken by the Hubble Space Telescope, 1995. Newborn stars form within gaseous globules in the Eagle Nebula in the Milky Way galaxy 7,000 light-years from Earth.

pp. 6–7. Artist's concept of a lunar base, projected for about 2020. Astronauts on the Moon test the hardware and new technologies necessary for deep spaceflight. The spaceship in this painting by Pat Rawlings is an excursion vehicle for a future voyage to Mars.

pp. 8–9. Artist's vision of the deployment of the Hubble Space Telescope, 1990. Using a robot arm, astronauts inside the space shuttle Discovery raise the telescope from the payload bay.

pp. 10–11. Photograph of space shuttle, 1983. Moving on its mobile launch platform at one mile per hour, the spaceship approaches launchpad 39A.

pp. 16–17. Artist's view of the first expedition to Mars, 1954. Astronauts in Earth orbit prepare a fleet of spacecraft for the long voyage, an event anticipated in this early painting by Chesley Bonestell.

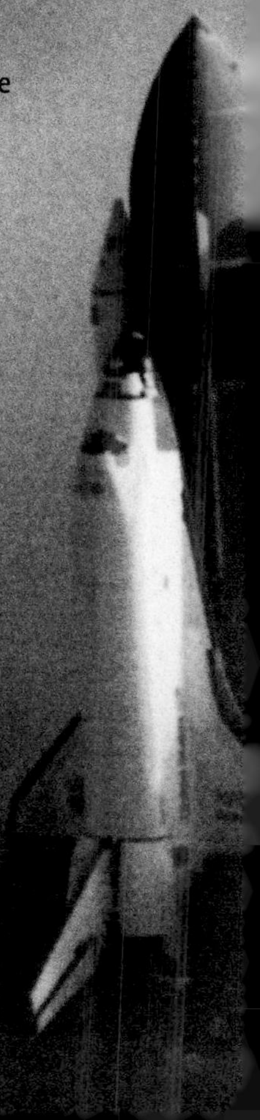

Contents

Foreword: Going A'journey

Ray Bradbury

When the Earth cooled and Life first quickened, the impulse to travel, move, and to survive was implicit. It has been given a new shape and drive by the impulses you will find in this remarkable book.

We don't know where we're going, but we're on our way, and the place that we're going to is so immense that it cannot be described. It is out there waiting. It is filled with billions of stars and billions of light-years of travel. We won't go very far, not far enough to ensure that our children will live forever, but their children's children's children will indeed live forever and beyond.

Friends ask, "Do you believe in Darwin?" I reply, "Yes, I do."

People ask, "What do you think of Lamarck?" I reply, "Lamarck is fine."

People ask, "What about Creation? Do you believe in the Old Testament?" I say, "Yes."

They cry, "How can you approve of all *three!?*"

"Because," I say, "nothing is proven."

I run what I call a religious, scientific delicatessen. When I wake each day I say, "Give me some of deese, some of does, some of dem." So I pick amidst wonders and live a compatible life.

★ ★ ★

Kerouac had his On The Road, Mao had his Long March, Lord Chesterfield wrote of Going A'journey, and now here we have a new kind of travel log.

This is the Second Book of Symbiosis.

The first was the Book of Dichotomy.

Let me simplify.

During the hundred thousand years prior to our age, we lived in an age of dichotomy, separate from the Universe, separated by our ignorance, separated by time and space.

Now, with the arrival of the people responsible for this book, we witness a symbiosis between ourselves and the cosmos.

★ ★ ★

Here, with *Imagining Space,* is a new book of Creation, a new New Testament. From this century on we will dedicate ourselves to the start of a special time; the beginning of the end of our life on Earth. It is the time of going away. In the next century we will know commencement toward Alpha Centauri.

Artist's rendering of a spacecraft returning from the Moon, about 2020.
A spaceship leaves a lunar base to journey homeward, its destination an Earth-orbiting space station.

It may take us a thousand years to landfall there, but it will be done.

The creators of this book are metaphorical cartographers who will chart the Universe and then move to fill that void with the minds and the flesh of mankind.

Nietzsche, in the midst of his movable feast of ignorance, declared that God, as far as he was concerned, was dead. God, as Cosmos, lived on, ignoring Nietzsche who, meanwhile, died.

Which leaves us in a new age where God is revived by the simple act of space travel. Our body, our blood, our spirit, and our mind are re-invigorated by the concept of touching the Universe. In the time ahead we will see a retrieval of faith as man's imagination moves from Earth.

With such star splendors on all sides, it would be difficult not to know the joy of Creation; not in the old Earth-bound religious sense, but in a true spirit of delight and discovery.

George Bernard Shaw imagined this in his plays; his exhilaration at the hidden impulse of humanity.

Years ago, I wrote GBS Mark V, in which I took an electro-mechanical robot Shaw into space so that I could rev him nightly to hear him spit forth wild genetic digital truths.

What were we doing in the Universe?

He cried, "Behold! We are Matter and Force making itself over into Imagination and Will. The flesh of the Universe which knows not itself rouses up in us, beholds the vast spread of the stars and says: I Think, therefore the Universe exists! The beasts that were for a short while at the mouth of the cave escape now from the cave to burgeon Mars, fire through the solar system and threshold the stars."

Indeed, too soon from the cave, too far from the stars. That changes, even as we watch. The stars are still distant, but we start the journey.

Do I say that new religions will be born on Earth, with new buildings, new congregations, new texts? No. I say something of greater size.

In my meetings with the Smithsonian years ago, they asked me to create a fresh planetarium program. In their auditorium I witnessed a show which induced snoring everywhere. The Smithsonian officials cried, "What are we doing wrong?"

I said, "My God, you're *teaching* when you should be *preaching!* Get back! Let me write 'The Great Shout of the Universe'; the passion of mysterious life flowering to inhabit interstellar light-years!"

They stood back. I wrote a thirty-two-minute show. They sent me twenty-eight pages of criticism. I phoned and said, "What are you trying to do? Go back to boring people? You can't shine the textbook on the ceiling. What you must burn up there is the glory of the amazing life of people on Earth discovering themselves and wanting to flee to the Moon! What bothers you most about my script?"

They said, "You have the Big Bang happening 10 billion years ago."

I replied, "What year *did* the Big Bang occur?"

"*Twelve* billion years ago."

"*Prove* it," I said.

Well, that ruined my Smithsonian connection.

I saw that they wanted science and I wanted creative tantrums.

I phoned and said, "How much do you owe me?"

"$15,000."

I said, "Give me $7,000 and let me go. This is a bad marriage."

I went away from there with the feeling that I sensed more about the cosmos than they did.

I thought, If the Universe stretches a billion light-years in this direction, that direction, and yet another direction, why couldn't it be that instead of the Big Bang, the stars have been here *forever*?

Impossible? But what is more impossible than the Big Bang? Choosing between the two I decided that the Universe was always here. If that's true, why don't we relax and accept our role in the Cosmos as beholders and celebrators.

Which returns us to Shaw, who saw mankind as this creature with notions to move on, like the old song, "We don't know where we're going, but we're on our way." The seed was always there.

The very soil beneath us once had a subsoil of dreams, inert matter that "longed" to be free, but only waited for an impulse of lightning struck and restruck in impulses over millions of years until at last the hidden dream spoke and said, "Alright, I give up the ghost." And out of primitive thought and motion and sustenance, it came forth on Earth; we know not where or when, but we inherit the why. We have arrived to audience the wonders, to speak the miracles. We are shadow and substance; shadow which cloaks itself in substance and calls itself Life.

★ ★ ★

In my garden not long ago I encountered an immense golden spider that had taken a vast space and cast across it an incredible cathedral window of web. I would not for any reason break that amazing window, for it encapsulated mankind and our inner dream and invisible outreach toward space, time and forever.

In that tiny brain cell was all the knowledge inherited to spin a mighty architecture on the summer air. Untaught and untouched by the countless generations of similar golden creatures, it moved to capture the air and insure its sustenance and survival.

We do the same.

Not quite knowing. We know not why we thread an architecture of travel in a fiery path across a winter space and warm far worlds with our breath.

Preface

Fifty years ago, a group of visionaries issued plans for the exploration of space. At the time, no humans had ever flown beyond the stratosphere. No artificial satellites orbited the globe. More people believed in atomic-powered railroad trains than in spaceflight. In a 1949 Gallup poll, only 15 percent of the respondents agreed that humans would walk on the Moon by the year 2000.

People with visions of space exploration worked to dispel that skepticism. They published books; they wrote articles. In 1949, a fantastic book appeared. On the dust jacket, a slender rocketship sat on a lunar plain. Humans in spacesuits explored a well-eroded terrain. Inside the book, a series of illustrations by Hollywood special effects artist Chesley Bonestell showed how the Moon and Mars, Saturn, and the other planets might appear to the first explorers to reach them. Willy Ley, a German refugee, wrote the text. They called it *The Conquest of Space*.

Two years later, in 1951, Arthur C. Clarke laid out a series of steps that he said humans would follow as they began to explore the unknown. Head of the British Interplanetary Society, this young novelist wrote a nonfiction book titled *The Exploration of Space*. Eight years later, leaders of the U.S. National Aeronautics and Space Administration (NASA) adopted as their long-range plan a set of missions nearly identical to those Clarke had proposed.

In March 1952, readers saw a winged rocketship ascending into space on the cover of *Collier's* magazine. The magazine contained an article by Wernher von Braun, technical director of the German rocket works during World War II and subsequently a leader of the U.S. Army rocket effort, describing how the United States would build an Earth-orbiting space station serviced by winged spaceships. "What you will read here is not science fiction," the editors of *Collier's* magazine announced. "It is serious fact."

People like von Braun, Bonestell, Clarke, and Ley promoted a vision of space exploration that came to dominate both popular culture and aeronautical engineering. Walt Disney summarized the vision in three widely viewed television programs and employed it as the centerpiece for the Tomorrowland section of his Disneyland theme park. A new generation of science fiction writers described Martian colonies (Ray Bradbury), launched starships (Gene Roddenberry), explained robotics and the foundation of galactic empires (Isaac Asimov), and forecast what would happen by 2001 (Arthur Clarke and filmmaker Stanley Kubrick).

The vision that such people proclaimed became, in the words of historian Frederick I. Ordway, America's "blueprint for space." Their forecasts reappeared in NASA's first long-range plan (1959), the report of the president's Space Task Group (1969), and the recommendations of the National Commission on Space (1986).

Much of the vision came true during the twentieth century. In one of the *Collier's* articles, von Braun predicted that humans would reach the Moon by 1977. In reality, they beat that prediction by eight years.

In demonstrating how humans would fly from Earth into space, von Braun described a reusable space shuttle, admitting "this will be an expensive proposition." It was more expensive than anyone anticipated. The inability of rocket scientists to reduce the high cost of spaceflight cast a long shadow across a variety of activities, from efforts to build orbiting space stations to the launching of planetary probes.

As in any new endeavor, surprises arose. Among the most unexpected has been the power of remote sensing. Few early pioneers envisioned the ease with which engineers could transmit images and data from spacecraft back to humans on Earth. The problem of communicating with distant spacecraft so perplexed Robert H. Goddard, the American rocket pioneer, that he suggested placing ignitable flash powder on the first rockets to the Moon so that humans would know they had arrived. Members of the British Interplanetary Society suggested flashing lights.

Advances in remote sensing allowed scientists to build spy satellites, investigate the outer planets, and retrieve photographs from the surface of Mars. Remote sensing helped the United States win the Cold War and provided images of the solar system as stunning as those painted fifty years ago by Chesley Bonestell.

New pioneers will fill the next fifty years with similar achievements, disappointments, and surprises. This volume celebrates the fiftieth anniversary of the visions promoted by von Braun, Bonestell, Clarke, Bradbury, Ley, and others at the midpoint of the twentieth century, and anticipates the amazing events potentially achievable within the next fifty years.

***The Conquest of Space*, 1949.**
Astronauts disembark from a modified V-2 rocket in Chesley Bonestell's portrayal of the first expedition to the Moon. The book laid out a realistic program of space exploration for the following fifty years.

01 : Wondrous Visions

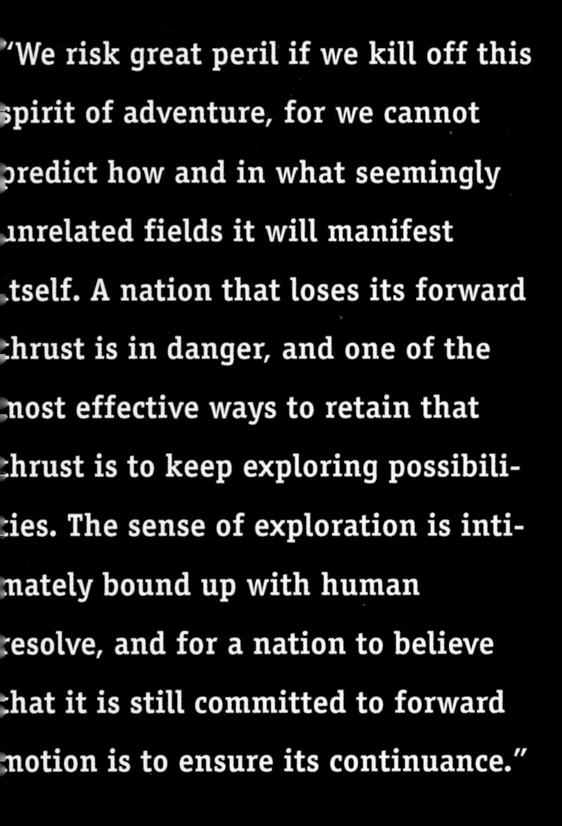

"We risk great peril if we kill off this spirit of adventure, for we cannot predict how and in what seemingly unrelated fields it will manifest itself. A nation that loses its forward thrust is in danger, and one of the most effective ways to retain that thrust is to keep exploring possibilities. The sense of exploration is intimately bound up with human resolve, and for a nation to believe that it is still committed to forward motion is to ensure its continuance."

—James A. Michener, author, 1979.

* * * * * *

The first generation of people to explore space

grew up with the most wondrous visions of what they might achieve.

They watched humans walk across a barren and craggy moon, nineteen years before Americans actually arrived, in George Pal's 1950 movie *Destination Moon.*

They saw winged spaceships fly off the covers of books and magazines, thirty years and more before the first space shuttle flew.

The first generation scanned *Life* magazine and books like *Conquest of Space* for paintings of astronomical landscapes that showed how other planets and moons might appear, long before the first NASA spacecraft arrived at Jupiter and Mars.

Scientists assured the public that life-forms existed on other worlds, certainly on Mars, and the first generation watched a succession of Hollywood films featuring strange, alien beings.

The first generation witnessed a large, rotating space station in *2001: A Space Odyssey,* a movie released in 1968, the year before Americans first landed on the Moon. Two enormous wheels turned slowly above Earth, providing a transfer point to lunar bases and beyond. The United States began to build its first large space station thirty years later.

Children of the Cold War, the first generation heard experts proclaim the belief that the nation that controlled space would control the world. Americans worried that Russians might get there first, and wondered whether missiles and nuclear bomb tests had attracted the attention of extraterrestrial civilizations. In a classic 1951 movie, *The Day the Earth Stood Still,* the alien captain of a flying saucer lands on the Ellipse in front of the White House and rescues the world from the madness of nuclear proliferation.

To the average American who grew up during the first half of the twentieth century, space travel was "that Buck Rogers stuff," futuristic and unreal. Buck Rogers was a comic strip character who as of 1950 had appeared in daily newspapers for more than twenty years. The fictional character lived in a world of spaceships and death rays five hundred years in the future.

Fifty years ago, in the middle of the twentieth century, a group of enthusiasts set out to convince people raised on Buck Rogers and other such fantasies that space exploration was real. Space, they said, was the next frontier. Space exploration would change human civilization in ways that people could hardly imagine. Humans would learn to live and work in space, just as they had migrated across Earth's continents and oceans. They would build spaceships and space stations and land humans on the Moon. They would establish lunar bases and Martian colonies and discover extraterrestrial life. "Humans will

preceding pages

The vision and the reality, 1952 and 1971.

Wernher von Braun's plans for lunar exploration involved fifty astronauts and three large spacecraft, painted by Chesley Bonestell for a 1952 issue of *Collier's* (left). Just nineteen years later, astronaut James B. Irwin stands in front of the lunar module Falcon that he and David Scott piloted to the Hadley-Apennine landing site during the Apollo 15 mission (right).

opposite

Photograph of Earth and Moon, 1992.

When Chesley Bonestell and Willy Ley published *The Conquest of Space* in 1949, only 15 percent of Americans believed that humans would reach the Moon by 2000; we achieved the goal in 1969. Twenty-three years later on its way to Jupiter, the Galileo spacecraft looked over its shoulder and took this exquisite photo of Earth and its satellite.

conquer space soon," the March 22, 1952, cover of *Collier's* magazine proclaimed. Nine years later, in 1961, Soviet cosmonaut Yuri Gagarin became the first human to fly in space, fulfilling the first step in the Buck Rogers dream.

Collier's, March 22, 1952.
A series of articles in *Collier's* between 1952 and 1954 generated public excitement about the possibility of a wondrous future in space and helped establish the sequence of exploration.

The Potential of Space The first people to write seriously about space exploration anticipated most of the uses to which this new realm would be put. Before actual flights began, they foresaw the commercial potential of space. In 1945 a young radar instructor named Arthur C. Clarke described the way in which satellites placed in geostationary orbits could be used to create a worldwide communications network. Clarke went on to become a famous science fiction writer, preparing the screenplay for *2001: A Space Odyssey*. The concept of communications satellites transformed the world.

Rocket scientists realized that space would become the high ground for military operations in the modern era. They warned that hostile nations might place nuclear missiles in space or launch them toward Earth from lunar bases. Wernher von Braun, the most outspoken advocate for space travel at that time, predicted that military leaders from space-faring nations would build orbiting reconnaissance platforms to spy on each other. "It will be almost impossible for any nation to hide warlike preparations for any length of time," von Braun prophesied in the March 22, 1952, *Collier's* magazine. The editors at *Collier's* were so impressed with this argument that they urged U.S. leaders to begin work immediately on a station from which such reconnaissance activities could take place. "In the hands of the West," the editors said, such platforms "would be the greatest hope for peace the world has ever known. . . . It would be the end of Iron Curtains everywhere."

Fears of nuclear weapons falling like bombs from space were groundless. As a means of nuclear deterrence, the surface of Earth and its seas provide far better platforms for launching guided missiles than orbital bases, a fact that led the superpowers in 1967 to sign a treaty banning nuclear weapons from space. Reconnaissance satellites, on the other hand, fulfilled the potential predicted for them. They played a decisive role in the outcome of the Cold War, piercing the Iron Curtain, providing the United States with an effective early warning system for detecting missile launches, and permitting the verification of arms control agreements.

Space enthusiasts foresaw the advantages that space provides for observing Earth and assisting with navigation. They correctly predicted that orbital platforms would forecast the weather and fix the position of travelers below.

Astronomers yearned to place telescopes in space, above the atmosphere that otherwise obstructed their view. Von Braun, who favored human spaceflight, nonetheless recognized that such observatories would have to be run without humans onboard. "The movements of an operator would disturb the alignment," he correctly observed. Experts like von Braun correctly anticipated the importance of space observatories in advancing scientific understanding, although they missed one important technical point. Von Braun believed that astronauts in space suits would be needed to insert and retrieve film; hence, he proposed that telescopes be placed in orbits near space stations with humans onboard. In practice, humans are needed to service and upgrade space observatories, but they are not required to change the film.

Space enthusiasts believed that exploration would provide technological solutions to human problems, eventually assuring the survival of humankind. As a result of space exploration, energy shortages would disappear. Hermann Oberth, an early rocketeer, proposed the construction of orbital reflectors to direct solar radiation to energy-starved regions on Earth. The problem of excess population might also fade away. Jack Williamson, writing a series of science fiction stories under the pseudonym Will Stewart, proposed methods for attaching atmospheres to otherwise lifeless bodies, making them suitable for human habitation. Space technology could also be used to make habitable the undeveloped portions of Earth. "It would be far easier to make the Antarctic bloom," Arthur Clarke prophesized, "than to establish large, self-supporting colonies on such worlds as Mars, Ganymede or Titan." Yet a day would come, Clarke predicted, when humans would look ravenously at other planets as places to relieve an overcrowded world.

***Collier's*, April 30, 1954.**
For the early pioneers, the ultimate objective was a human expedition to Mars, depicted here in the landing phase by Chesley Bonestell.

Indeed, migration into space in the distant future might provide humans with a form of immortality. Robert Goddard, the American rocket pioneer, proposed that humans send the genetic material necessary to establish life on other planets on long voyages across intergalactic space. "Although a discussion such as this may seem academic in the extreme," he speculated in 1918 and again in 1943, "it nevertheless poses a problem that will some day face our race

Wernher von Braun and Willy Ley, 1954. Von Braun (left) and Ley (right) meet on the set of a Walt Disney television program that helped build public support for an ambitious space program.

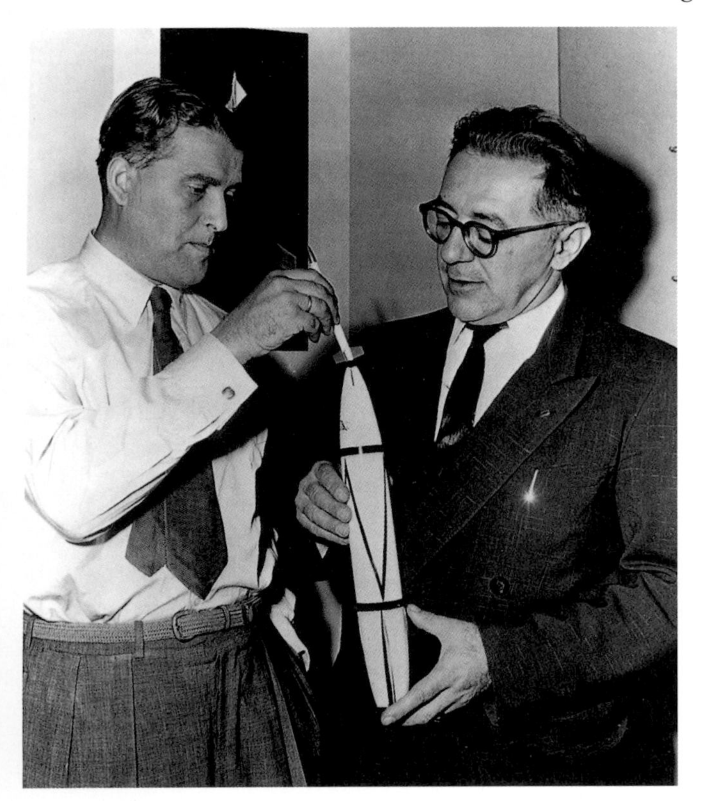

as the sun grows colder." Life in the form of "granular protoplasm" could travel to all parts of the Milky Way. Once established, it might evolve as it had done on Earth. Goddard urged the designers of these expeditions to send "as much as possible of all human knowledge, in as light, condensed, and indestructible a form as possible, so that the new civilization can begin where the old ended."

Commerce, national security, Earth observation, science, and human survival all motivated the pioneers of space exploration. These were important reasons for going into space, but not the only ones. Many of the first group of pioneers wanted to explore space simply for the challenge involved. They were curious. They wanted to go out and see for themselves what was there, armed with little more than their belief that knowledge of the unknown was on the whole beneficial for humankind.

The von Braun Paradigm The exploration of space, in the minds of early pioneers, began with rocketry. First came small rockets. Shooting a rocket straight up into space is a relatively simple affair, as technicians learned in 1949 when they fired an Army WAC Corporal rocket, mounted on top of a German V-2, to a record height of 250 miles, the altitude at which the space shuttle commonly flies. The rocket went straight up, stopped, and came back down. To orbit Earth, a rocket needs to retain a lateral speed (parallel to the surface of Earth) of approximately 18,000 miles per hour. At that point, its outward velocity balances gravitational pull, and it remains in orbit.

To achieve such speeds, rocket scientists mounted rockets on top of rockets, complicating the operation to a considerable degree. In 1958, members of the Army rocket team led by Wernher von Braun mounted a 30-pound satellite on top of a cluster of solid fuel rockets, placing them on top of a Jupiter-C liquid fuel rocket, a direct descendant of the German V-2. Team members aimed the rockets perfectly, fired them sequentially, and shot the first U.S. satellite into orbit around the world.

To fully explore space required much larger rockets. Well before he launched the first Earth satellite in 1958, von Braun presented plans for a massive, three-stage rocketship that could propel humans and associated equipment into space. In 1962, NASA executives told von Braun to start building the 364-foot-tall Saturn V, the three-stage rocket that launched Americans to the Moon.

In von Braun's original plan, the three-stage rocketship propelled an airplane-shaped shuttle into space. In 1972 government officials authorized workers at NASA's Marshall Space Flight Center, which von Braun had organized,

Artist's depiction of the first expedition to the Moon, 1952. In a series of Chesley Bonestell paintings prepared for *Collier's*, three large spaceships rendezvous in Earth orbit (top). After landing on the Moon, astronauts lower a section of their cargo ship for use as a habitat (center). Traveling in pressurized rovers, they mount an expedition to Crater Harpalus, 250 miles away (bottom).

Artistic depictions of space stations.
In 1952, Wernher von Braun proposed a rotating space station with artificial gravity as a jumping-off point for expeditions (right). In 1994, NASA began design work on a nonrotating space station that could be used as an orbiting laboratory for microgravity research. Assembly of the International Space Station began in 1998 (below).

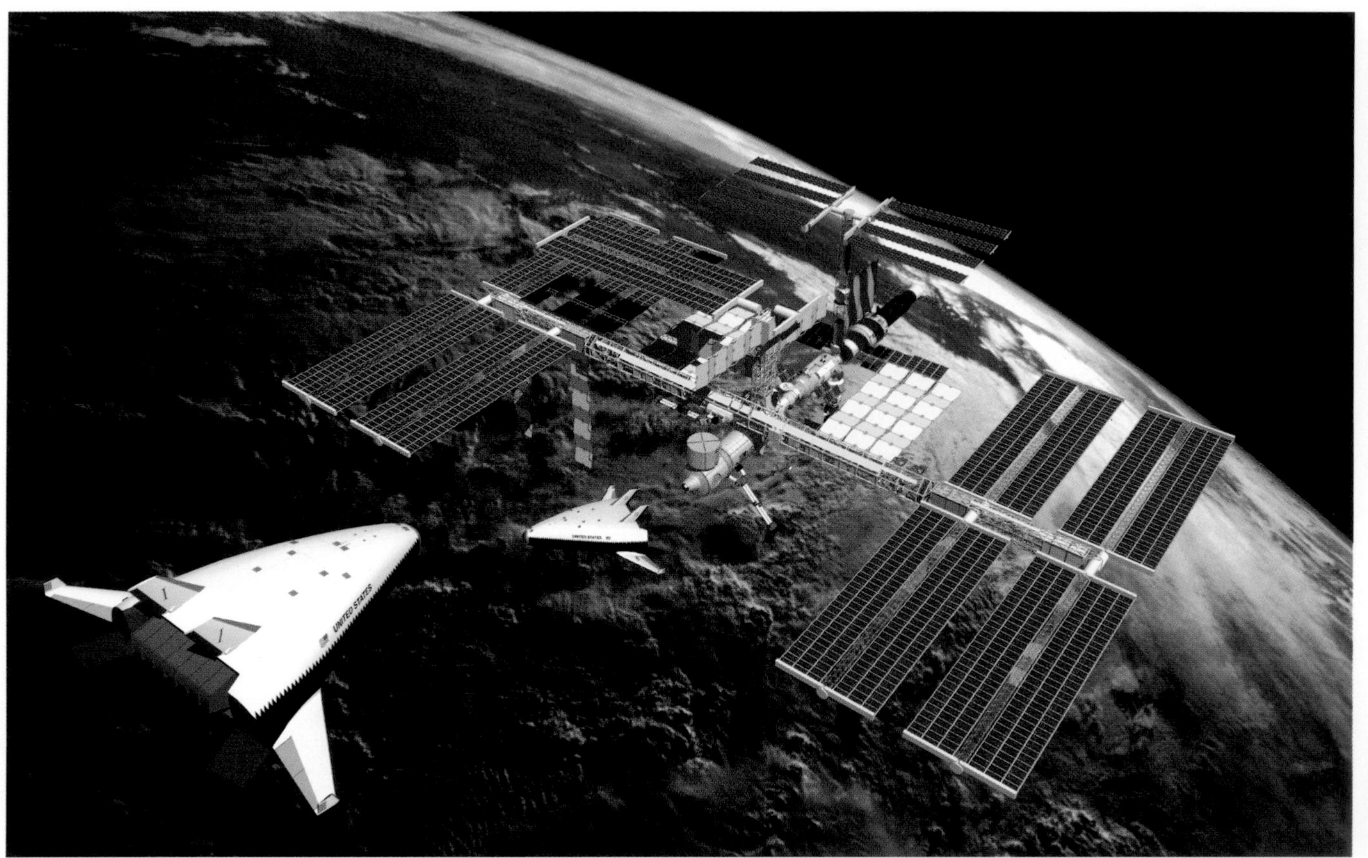

to begin work on the propulsion system for a winged space shuttle. Two slender solid-fuel rocket boosters, augmented by three hydrogen-burning engines, generated all of the thrust necessary to launch a 200,000-pound orbiter spacecraft and cargo weighing up to 50,000 pounds.

All of the early pioneers wanted to start with an orbital space station. Again, von Braun took the lead. He proposed construction of a 250-foot-diameter wheel, assembled out of twenty prefabricated sections, which when linked together could hold an eighty-person crew. Von Braun's wheel-shaped design provided the inspiration for the imaginary space station in the science fiction movie *2001: A Space Odyssey*. That station contained two wheels, each 900 feet across. The interiors for von Braun's design resembled the innards of a submarine and held a maze of observation posts, elevator cages, and equipment rooms. Film directors altered this design, creating a spacious interior that resembled a modern hotel. The set directors placed a sign at the registration desk announcing that the Hilton Corporation managed the station interior.

Having established a base in low Earth orbit, humans would be ready to explore the Moon. As the principal spokesperson for the space exploration movement in America, von Braun envisioned an expedition much like that conducted by Meriwether Lewis and William Clark on the American frontier. In a 1952 plan, outlined in *Collier's* magazine, von Braun described a fifty-person expedition on a six-week reconnaissance of the Moon. Technicians in space suits, von Braun proposed, would assemble three very large spaceships in the vicinity of an orbiting space station. Each spaceship would measure 160 feet in length and, fully fueled, weigh more than 4,000 tons. Two of the ships would carry sufficient fuel to land on the Moon and return to the Earth orbiting station. The third would carry a 75-foot-long cargo container to the lunar landing zone. Once on the lunar surface, astronauts would unload supplies from the cargo container using construction cranes. The empty cargo container would be removed from the landing craft and split in half to create two ready-to-use Quonset huts for the expedition's base camp.

Von Braun proposed that astronauts set up their base camp in a crevice beneath a towering mountain range, to protect the expedition from cosmic radiation. To do this, astronauts would need to tow their equipment from the landing site to the base camp using three pressurized tractors. "The principal aim of our expedition during this first lunar exploration will be strictly scientific," von Braun and astronomer Fred Whipple promised, by which they meant that military objectives would not dominate the mission. (At the time he outlined his proposal, von Braun designed missiles for the U.S. Army.) Expedition leaders would probe the origins of the Moon, conduct experiments, and search for raw materials. They would dispatch ten people on a ten-day round-trip excursion

following pages

Mars, real and imagined.
In 1954 von Braun and Ley proposed that astronauts hovering in orbit around Mars attach wings to interplanetary transit craft (top left). The spaceships would skid to a landing (bottom) and humans would establish a Martian base (top center, left). The first vehicle that actually landed on Mars was the 1976 robotic Viking lander (top center, right). This artist's rendering shows the lander, still encased in its cone-shaped aeroshield, separating from its orbiting spacecraft and preparing to descend to the surface. The surface of Mars—a cold, desolate place—was photographed by the 1997 Mars Pathfinder (top right).

to Crater Harpalus, 250 miles away, proceeding in a convoy of tractors and trailers. Plans called for the expeditionary corps to remain on the Moon for forty-two days.

Repeated experience with lunar exploration would prepare humans for what nearly all of the early visionaries believed to be the next logical step: a voyage to the planet Mars. Very little was known about Mars at that time; most experts believed that human visitation would be required before all of the planet's secrets were revealed. In 1948, von Braun developed specifications for a Mars expedition that he hoped to present as part of a science fiction novel. The novel was never published, but the technical plans were.

Speaking again for the advocates of spaceflight, von Braun suggested that the first expeditionary crew travel to Mars in a flotilla consisting of ten spaceships. Upon arrival, the flotilla would orbit Mars. From their orbital base, astronauts would descend to the surface in three airplanelike spacecraft. Von Braun proposed to land the first craft on the polar ice cap, the only surface thought sufficiently smooth for a safe landing, using skis instead of wheels. After unloading tractors and supplies, the crew would drive 4,000 miles to the Martian equator and prepare a landing strip for the other two planes. For one of von Braun's books on the exploration of Mars, Chesley Bonestell painted a famous landscape incorporating the winged spaceplanes and the expeditionary corps surveying a desertlike terrain.

The expedition would remain on Mars fifteen months, waiting for Mars and Earth to realign themselves properly for the return voyage. Removing the wings from their landing craft, ground crews would set the spaceplanes on their tails. The expedition team would gather onboard, blast off, rendezvous with the spaceships in which they had come, and head home.

Von Braun believed that the space station could be fully operational by 1967. He set 1977 as the date for the first lunar landing. Regarding the ability of humans to venture to Mars, he was more cautious. "It will be a century or more," von Braun predicted. At the midpoint of the twentieth century, most experts believed that humans needed to gather a great deal of experience in low Earth orbit and on the Moon before venturing afar.

Von Braun's timetable took a strange turn once actual missions began. Recognizing that the United States could not beat the Soviet Union in establishing the first orbiting space station, President John F. Kennedy chose to race to the Moon. As part of the crash program to put Americans on the Moon, lawmakers increased NASA's budget tenfold. The sudden infusion of money encouraged space enthusiasts to advance von Braun's timetable. Thomas Paine, NASA Administrator when Americans landed on the Moon, believed that spending rates of Moon-race magnitude would allow the United States to venture toward Mars far in advance of von Braun's hundred-year goal. Paine recommended a target date of 1986. Optimism proved so contagious that Stanley Kubrick, producing his science fiction film at the height of the Moon race, predicted that the United States would launch an expedition to Jupiter by 2001. According to space historian Frederick Ordway, who served as a technical adviser on the film, Kubrick worried that he had underestimated accomplishments that could be achieved by that time.

Alas, the optimists failed. Congress cut spending for space exploration substantially as the Apollo Moon program wound down. By the end of the twentieth century, humans had stood on the Moon, had built astronomical observatories, and had begun constructing a large space station. They had not, however, built any lunar bases or dispatched a human expedition to Mars.

The Economics of Space Exploration In one respect, Paine and fellow space enthusiasts saw the future clearly. Paine believed that the pace of space exploration depended not just on advances in technology, but also on the size of one's economy. Space exploration is expensive, something that societies with small economies can hardly afford. As economies expand, so does the capacity for space activities, a thesis Paine expressed as head of the 1986 National Commission on Space.

Paine hoped that NASA's budget would grow along with the United States economy. If the Congress simply allocated a fixed share of the gross domestic product to space exploration, NASA spending would increase threefold in the next fifty years in real dollars (adjusted for inflation). This would raise government support for space exploration beyond the record levels set during the Moon-race years. It would permit a substantial increase in government-funded space activities, especially as the cost of individual missions declines due to advances in microelectronic technology and rocket propulsion.

NASA space activities, however, are just the tip of the wedge. The groundbreaking missions conducted by NASA officials open pathways on which a phalanx of merchants, scientists, soldiers, entrepreneurs, and citizens from throughout the world can travel. NASA provides the expeditionary corps; many people follow behind.

Military space expenditures in the United States have matched NASA spending dollar for dollar since 1980 and will continue to rise in the twenty-first century. Economic growth has propelled other nations into the space-faring club. Sixteen governments, led by the United States, are cooperating to build the International Space Station. In Europe, China, and Japan, government officials have developed their own rocket and space programs. Other governments now spend one dollar for every two dollars that the United States government allocates

"To see the Earth as it truly is, small and blue and beautiful in that eternal silence where it floats, is to see ourselves as riders on the Earth together, brothers on that bright loveliness in the eternal cold—brothers who know now that they are truly brothers."

—Archibald MacLeish, poet, 1968.

to its space activities. Worldwide, as of 2000, all governments spent about $40 billion on civil and military applications. As a proportion of that total, NASA spent $14 billion.

The most dramatic growth in space spending will not arise from government treasuries, however. It will come from the private market. The year 1998 proved a landmark in that regard: according to the best estimates, commercial space revenues exceeded government spending worldwide for the first time. Space entrepreneurs raised more money through private sources than governments allocated through tax revenues. Analysts estimate that commercial space revenues will expand fourfold during the first decade of the twenty-first century and continue to grow thereafter.

A great deal of money will be available to finance space activities in the twenty-first century. Total worldwide expenditures on space could surpass $500 billion, in the value of year 2000 money, before the midpoint of the twenty-first century. The share provided through civil space agencies like NASA will decline, reversing the situation that existed at the beginning of the space age when government expenditures dominated the whole. The influence of public expenditures, however, will remain significant, opening the way for other activities to follow.

The vision of spacecraft plodding toward the Moon and planets in massive, tax-supported flotillas has been an entertaining one; it inspired a half century of exploration. It is unlikely to represent the whole trajectory of future space activities, however. Some of the original vision will come true, but much will be supplanted by accomplishments that the original visionaries never imagined.

Photograph of space shuttle Challenger, 1983.
Nearly all the early visionaries viewed the development of reusable launch vehicles as a necessary step toward reducing the high cost of spaceflight. This image was taken by a free-flying robot deployed from the orbiter's cargo bay.

02 : The Search for Extraterrestrial Life

preceding pages
Images of extraterrestrials.
In 1908, eleven years after the publication of *War of the Worlds*, an artist envisioned how intelligent creatures evolving on low-gravity Mars might appear (left). In 1996, scientists released photographs of wormlike structures found in a meteorite that came from Mars, possibly extraterrestrial fossils (right).

Humans will soon possess the tools necessary to answer one of history's most penetrating questions: are we alone or do we coexist with living creatures on other spheres? In the next fifty years, humans will search for creatures within our solar system and look for signs of life on planets orbiting nearby stars. If life is out there, it will be found.

Most scientists believe that simple life-forms—perhaps like bacteria—exist on other worlds. More than one-third of the American public believes that creatures "somewhat like ourselves" reside on other planets, a belief reinforced by a plethora of films and novels anticipating the shape of alien beings. In a 1975 *Scientific American* article, astronomers Carl Sagan and Frank Drake estimated the total number of civilizations in the Milky Way galaxy "at or beyond the earth's present level of technological development." They set their estimate at 1 million civilizations. They derived their million-civilization prediction from a simple formula based on estimates of the number of solar systems in the galaxy, the number of planets suitable for life, and the fraction of suitable planets where life can evolve into complex forms.

No one on Earth produced direct proof of alien life during the twentieth century, a matter of considerable disappointment to people engaged in the search. The wholly circumstantial nature of the evidence, however, has not dimmed enthusiasm for the belief that humans will discover extraterrestrial life if they just look hard enough.

The Hunt for ET Belief in otherworldly beings is deeply seated, part of the human experience. It satisfies an old and pressing need. By contemplating contact with such creatures, humans grant themselves a privileged position in the cosmos, worthy of visitation by God-like beings. In medieval times, people absorbed by religious doctrine believed that angels and demons inhabited the heavens, took human form, and visited Earth. In the spiritually conscious nineteenth century, people turned their attention to ghosts and sought methods for communicating with the dead. As science replaced superstition, people embraced extraterrestrials, especially those with advanced technologies. Belief in supernatural beings possessing superior powers has been part of human civilization for a very long time.

By exploring the Earth, the leaders of terrestrial expeditions inadvertently encouraged the belief in alien life forms. In centuries past, explorers returned from expeditions with tales of wondrous beasts: some mythical, such as Sir John Mandeville's fourteenth-century report of a race whose faces appeared on their chests, and others real, such as Charles Darwin's stories about lizards that swam out to sea. Given this tradition, people who turned their attention toward the heavens naturally anticipated similarly strange creatures on distant worlds.

Religious history also encouraged scientists to embrace the notion of a "plurality of worlds," the doctrine that living creatures exist on many spheres. For centuries, scientists have fought religious institutions wedded to the anthropocentric belief that God fashioned humans in his own image and placed Earth at the center of creation. The Italian philosopher Giordano Bruno was burned at the stake in 1600 for suggesting that nature had filled the universe with a variety of inhabited worlds. Galileo Galilei was threatened with torture and placed under house arrest for advocating the doctrine that many planets, including Earth, circled the Sun. Charles Darwin was attacked by religious leaders for suggesting that humans evolved from lower life-forms. Scientists who have fought religious institutions over the physical location of Earth or the nature of evolution have been drawn naturally to the doctrine that Earth is neither physically nor biologically unique.

Early telescopes provided tantalizing views of other worlds. The images confirmed the existence of Earth-like bodies, but provided so few details that imagination led to inaccurate conclusions. Seventeenth-century astronomer and mathematician Johannes Kepler suggested that craters on the Moon were in fact city walls into which local inhabitants had constructed their homes. He presented this startling proposition in a science fiction story, but the proposition was repeated as fact two centuries later by a Munich astronomer Franz von Paula Gruithuisen, who claimed to see walled cities on the Moon. Belief in the habitability of the Moon continued well into the nineteenth century, when telescope technology improved sufficiently to confirm it to be a dead and airless world. Speculation consequently shifted to Venus and Mars, aided by a theory proposed by Pierre-Simon de Laplace, a nineteenth-century French astronomer. Laplace suggested that the mechanics of rotation had caused the planets to coalesce sequentially. According to this history of the inner solar system, Mars formed first, followed by Earth and then Venus. As the oldest of the inner planets, Mars represented what Earth would become in the distant future when its water and atmosphere began to disappear, while Venus remained a cloud-shrouded primordial world.

In 1962, NASA flew its Mariner 2 spacecraft past Venus, the first human-built machine to fly by another planet. The data from that and a succession of spacecraft confirmed the worst suspicions. Venus is a hellish place, with surface temperatures approaching 900 degrees Fahrenheit. The clouds thought to shroud carboniferous swamps contain sulfuric acid. The atmosphere, mainly carbon dioxide, is oppressively thick. No giant ferns or primitive animals grace the surface of this inhospitable world.

"For millennia humans have looked to the sky and wondered whether anyone else is out there looking back. We used to ask the philosophers and priests to answer this question according to their belief systems. . . . Today scientists and engineers can use our existing technology to try to answer that ancient question by doing experiments. Although we've been working at it fairly hard for four decades now, with null results, in truth we've hardly begun to search. . . . The important thing is to keep searching. The answer calibrates our place in the universe."

—Jill Tarter, SETI project scientist, 2000.

Fourteen years later, in 1976, NASA placed two Viking spacecraft on the surface of Mars. Again, actual observations deflated popular expectations. Cameras on the two spacecraft produced high-resolution photographs of the Martian landscape. The cameras operated in a peculiar fashion, recording one thin line at a time. This worried partisans of life on Mars. Any creature dashing in front of the spacecraft would appear as a break in a single vertical line. Without night lights, nocturnal animals could not be photographed at all. No creatures appeared, however. Even more disappointing, the cameras failed to locate lichen or moss on nearby rocks.

Each of the two Viking landers carried an automated biology instrument with a variety of test chambers capable of detecting microbes in the Martian soil. A special arm on each Viking lander scooped up soil samples and deposited them in the chambers. The machine added fertilizer, while instruments waited to record the gases that living plants or animals would produce. In another test chamber, automated machinery baked the soil, producing vapors that a gas chromatograph and a mass spectrometer could identify. The instruments worked. They detected trace amounts of the cleaning solvent that workers had used on the spacecraft before it left Earth and traces of what appeared to be hydrogen peroxide present in the Martian soil. To the disappointment of planetary scientists everywhere, however, the instruments detected no conclusive evidence of microbial life. The surface of Mars, the scientists grudgingly agreed, was probably sterile.

Science proceeds through fits of optimism and disappointment, rarely in a straight line. Initial results from Venus and Mars deflated expectations of a

Photograph of Mars, 1976. The two Viking spacecraft that landed on Mars in 1976 carried scientific instruments and cameras designed to detect surface life. Scientists found a dry, windswept planet inhospitable to the maintenance of surface-dwelling organisms.

lush and habitable solar system. Studies of life on Earth, on the other hand, encouraged scientists to expand the prospective number of sites where living creatures might establish a home. The two conclusions are not as contradictory as they seem. Simple life may begin under a wide range of conditions; complex life-forms such as plants and animals may be extremely rare.

How Did Life Start? For life to evolve, it must first arise. From the geological record, scientists know that life on Earth arose some 4 billion years ago, shortly after the planet cooled. Small organic molecules such as amino acids combined into larger molecules such as proteins, which in turn combined into droplets that acquired the capability of replicating themselves. This occurred before Earth developed an oxygen-rich atmosphere, under conditions that would be toxic to animal life.

Simple life, once it begins, is remarkably hardy. Streptococcus bacteria survived inside the television camera of the Surveyor 3 space probe after it landed on the Moon in April 1967. The bacteria endured nearly three years of radiation and temperatures only 20 degrees above absolute zero. Apollo 12 astronauts brought the camera back to Earth at the end of 1969, where the bacteria regenerated themselves.

Simple life thrived in Earth's oceans for 3.5 billion years. Five hundred million years ago, life on Earth underwent a remarkable transformation. For reasons not completely known, complex life-forms suddenly appeared. Soft-celled swimmers developed shells and backbones; creatures crawled out onto the land. The evolution of complex life-forms began during what scientists call the Cambrian explosion.

Artist's depiction of a robotic expedition to Mars, projected for about 2014.
To obtain Martian soil samples for analysis on Earth, a robotic rover collects material and delivers it to an automated spacecraft, seen here at liftoff for the flight home.

Observation of Earth suggests that microbial life appears as soon as suitable conditions arise. The range of conditions is remarkably wide. Bacterialike organisms exist in high-temperature geysers. They live miles below the surface of Earth in subterranean darkness. Living creatures thrive alongside thermal vents miles below the surface of the sea.

The frequency with which life begins is an important component in the formula for calculating the possible number of extraterrestrial homes. Scientists believe that life begins with remarkable ease; they are searching the solar system for supporting evidence.

By studying images taken from orbiting spacecraft, scientists know that water once flowed across Mars. Periodically, it may be flowing still. If life is ubiquitous, and liquid water persistent, life should have arisen on Mars. Scientists want to collect samples of Martian soil from ancient lakebeds and examine craters where water appears to have recently flowed. The samples could reveal evidence of now-dead life-forms or living microorganisms that may have retreated under the ground.

In ancient sedimentary rocks on Earth, marine plankton and early plant life leave biological markers called polycyclic aromatic hydrocarbons. Simple life-forms also deposit carbonate globules with elements like magnetite and iron sulfide around their rims. Samples of Martian soil may contain similar evidence. Some of those samples have already arrived. Close-up photographs confirm that comets and asteroids have bombarded Mars, as they have other bodies in the solar system. Ejecta from large impacts can achieve escape velocities, especially on a low-gravity body like Mars. About 16 million years ago, a large object struck Mars. Some of the ejecta escaped and drifted inward through the solar system until it encountered Earth's gravitational pull. After 15.9 million years in space, one of the rocks landed on an ice sheet in the Allen Hills region of Antarctica. An American scientist found the rock in 1984. Scientists knew that it came from Mars because gas trapped in small inclusions inside the rock exactly matched samples of the Martian atmosphere measured by the Viking landers.

Life on Mars The rock itself was more than 4 billion years old, produced at a time when Mars may have been warmer, wetter, and more hospitable to life. An investigating team from NASA compared the Martian rock with similar material taken from Earth. The results, to say the least, were astonishing. The rock contained polycyclic aromatic hydrocarbons and carbonate globules in proportions similar to those found at home.

In 1996, as the scientists prepared to publish their findings, news of the discovery leaked through higher government circles. President William

NASA

Clinton held a special news briefing. "If this discovery is confirmed," he proclaimed, "it will surely be one of the most stunning insights into our universe that science has ever uncovered." NASA administrator Daniel S. Goldin warned the media that "we are not talking about 'little green men.' These are extremely small, single-cell structures that somewhat resemble bacteria on Earth."

For their most dramatic, and least conclusive, bit of evidence, the research team produced pictures of strange-shaped objects inside the rock. They found the objects by chipping off pieces of the rock and photographing them with a high-resolution electron microscope. Could these be microfossils from another world? Skeptical scientists pointed out that the structures were one hundred times smaller than Earth-based bacteria. The structures might be minerals, skeptics said. The research team acknowledged that interpretation, but argued that the evidence, taken collectively, indicated the presence of life on Mars. Other scientists disagreed. The evidence for Martian bacteria, they responded, was not conclusive. Their fractious debate can be resolved in only one way. Humans must retrieve more samples from Mars.

Scientists would like to send spacecraft to Mars every two years, when the orbits of Earth and Mars converge. By 2014, scientists will have the capability to return samples from Mars, using robotic spacecraft sent on relatively low-cost missions. At each launch opportunity, scientists want to send an unfueled return vehicle together with a rover on a medium-sized rocket from Earth to Mars. The vehicle will land in an area of possible biological activity, such as an ancient lakebed. The rover will travel 6 miles or more from the landing site in search of samples. While the rover is out collecting samples, the return vehicle will collect gases from the Martian atmosphere and, utilizing a small onboard chemical plant, manufacture rocket propellants–probably methane and oxygen. The rover, on returning, will transfer its samples to sturdy containers on the return vehicle, which will blast off and return to Earth.

Unequivocal evidence of Martian life will be hard to find. Simple life flourished for 3.5 billion years on Earth before it evolved to the point where creatures with shells and skeletons left a visible fossil record. No visible fossils may exist on Mars. Analysis of samples returned from Mars may prove no more conclusive than the dissection of the Allen Hills meteorite. Spacecraft sent to Mars may discover their own bacteria, inadvertently brought from the Earth through a process called forward contamination. Full-scale investigation of Martian life processes may require a research station staffed by humans with a full range of scientific equipment.

If living organisms exist on Mars, scientists will want to know where they originated. They could have blown in on an asteroid. Cosmic dust contains hydrocarbons similar to those found inside the Allen Hills meteorite. Life on

Artist's rendering of a Martian research station, late twenty-first century.
Exploration advocates want to search for life on Mars. Technologies at a proposed facility include a pressurized rover, drilling rigs, tunneling devices, greenhouses, habitation modules, and a robotic airplane. The biologist-astronaut at lower right holds a Martian fossil chipped from the side of an ancient canyon near the equatorial Pavonis Mons volcano.

Earth and Mars may have had a common origin. Martian and earthly life-forms may share a common hereditary background, a possibility that can be investigated by comparing genetic codes. Perhaps life began on Mars, invaded Earth, and flourished. All earthly life may be Martian in origin—the possibilities are breathtaking.

Photograph of the surface of Europa, 1999.
Thera and Thrace are two dark, reddish regions of enigmatic terrain that disrupt older, ridged plains on Jupiter's moon Europa. Scientists suggest that liquid water from an ocean below Europa's icy shell has melted through the shell and that some form of life could reside in such a sea.

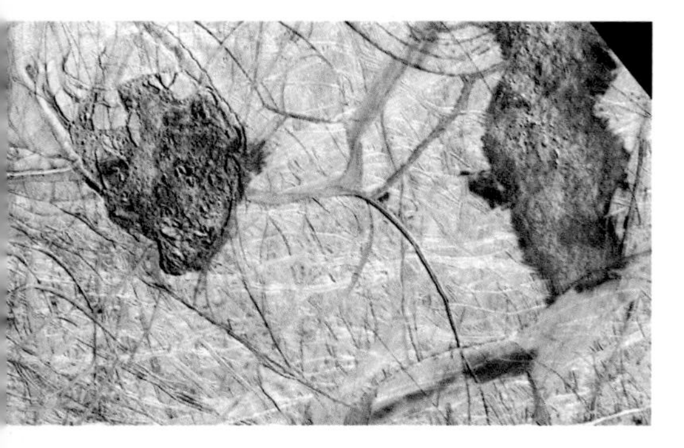

Europa and the Search for Life Scientists will look for life elsewhere in the solar system. One likely site is Europa, a satellite of Jupiter somewhat smaller than the Earth's Moon. Europa is covered with a crust of water ice, frozen by a bone-chilling surface temperature of minus 260 degrees Fahrenheit. The tidal tug-of-war created by the gravitational pull of Jupiter and its other moons creates an internal source of heat. (The same heating mechanism forms spectacular volcanoes on Io, another of Jupiter's large moons.) The heat may have created a subterranean ocean and a favorable environment for the development of life.

No sunlight falls on the subterranean seas of Europa. Surface ice blocks what little sunlight arrives. The absence of sunlight, however, need not have retarded the evolution of life. In the 1970s, marine scientists discovered dense populations of sea creatures at the bottom of Earth's oceans. Tubeworms, clams, and mussels lived well below the sunlight zone under conditions of forbidding hostility. The creatures cluster around hydrothermal vents, cracks in the ocean floor that provide a source of heat and nourishing chemicals. These creatures draw energy not from sunlight but from sulfur and carbon dioxide emitted through the vents.

The existence of life-forms thriving under conditions that resembled a toxic waste dump startled scientists. It caused them to widen their assumptions about the range of conditions under which life evolves. Scientists now believe that the so-called habitable zone includes hostile sites as far away as the moons of Jupiter and Saturn. If life can maintain itself at the bottom of the Earth's seas, it may have appeared under similar conditions on Europa.

A race is under way to discover the first evidence of extraterrestrial life. Some scientists favor Mars; others favor Europa. Scientists want to send space probes to Europa every four years, first to confirm the existence of liquid water beneath Europa's icy crust and then to search for life. The first probes will orbit Europa and look for evidence of liquid water oozing up between cracks in the crust. They will use radar to measure the thickness of Europa's ice. The surface of Europa is crisscrossed with fractures that show strange bands of color along fault lines. The colors may be caused by iron or sulfur or perhaps by organic compounds. Scientists want to know whether those materials contain compounds capable of supporting life.

If liquid water exists on Europa, scientists will want to search for crea-

tures within it. They hope to find simple bacteria; they will also look for more complex forms such as the tubeworms that exist along Earth's ocean floor. To search for life, scientists will send submersible robots to explore Europa's subterranean sea. This will be technically challenging, since the icy crust could be as much as 100 miles thick. Orbiting spacecraft scanning the surface of Europa can search for thin spots in the ice. Eventually a robotic spacecraft lands on one of those spots and unloads a cryobot—a robot capable of operating in icy cold or cryogenic conditions. The cryobot descends through the ice, trailing a communication line behind. A heat source, possibly a small nuclear reactor, allows the vehicle to melt its way downward. Once below the icy crust, the cryobot releases

Artist's depiction of the discovery of life on Europa, about 2025.
A cryobot has melted its way through the icy crust of Europa. Emerging below the ice, it releases a hydrobot, which searches for geothermal vents that may harbor life.

a submersible hydrobot–a robot designed to work under the sea. The hydrobot carries a lamp to expose the seabed and any creatures living on it. A camera and a variety of instruments survey the undersea. A communication line dispatches information from the cryobot to the surface of Europa and back to Earth.

If life is found on Mars or Europa, scientists will want to bring samples back to Earth. This could occur during the first half of the twenty-first century. Current plans call for samples to return directly to Earth, plunging into the atmosphere at high speed in the same way that Apollo astronauts returned from the Moon. Atmospheric friction would decelerate the precious cargo; parachutes would guide it to Earth where a crushable material such as Styrofoam would cushion its impact. Samples will be tightly encased in multilayer containers to protect them from earthly contamination. Once in research laboratories, they will be held in strict quarantine.

Some scientists would like to analyze alien samples on orbiting space stations, but there is no intrinsic advantage to this. All such stations eventually fall back to Earth. Given that fact, most scientists prefer to bring extraterrestrial samples directly to terrestrial laboratories.

If the samples contain bacterialike creatures, scientists will grow cultures. They will study how the microorganisms mutate before breaking the quarantine. In the hostile environment of their home planet, the organisms may be an endangered species. Scientists will work to keep them alive. The first live samples may be small, microbial, and dangerous. Alien microbes may view Earth as a very hospitable place on which to thrive.

The consequence of knowing that life begins–or began–on other bodies in the solar system will be profound. Most scientists believe that simple life-forms arise anytime a suitable source of energy enlivens organic chemicals in a liquid medium on a sphere of reasonable size. Scientists know that life exists on Earth in a wide habitable zone. They know that planets exist commonly around nearby stars. If life begins easily and thrives throughout a wide habitable zone and the number of candidate sites is very large, then humans live in a universe teeming with biological activity. The life-forms may be simple, but they may be everywhere. From knowing that life is everywhere, scientists are one step away from discovering the sites on which it has evolved into more complex forms.

Looking for Earth-like Worlds For centuries, astronomers and the public at large have known of only one planet in one solar system on which complex life resides: our own. Toward the end of the twentieth century, astronomers developed a method for identifying other solar systems. They did not observe other planets directly; rather, they deduced their existence by noting the behavior of the stars around which those planets travel.

As planets revolve around a star, they tug that object toward them, producing variations in the star's position that can be calculated precisely. The star also dims slightly when large planets pass in front of it. Using this knowledge, astronomers discovered more than two dozen planets in just five years. Planets—a key variable in the search for life—turn out to be quite common.

Finding Earth-size planets is more difficult. By the second decade of the twenty-first century, astronomers will be ready to launch a special space telescope capable of seeing planets otherwise hidden in the glare of their central stars. The technology is astonishing. Separate telescopes fly in formation using advanced computer techniques. The troughs of light waves from the central star reaching one telescope exactly match the crests of the light waves reaching another. The light waves cancel out, a process called "nulling." The glare from the star disappears, and planets suddenly appear. Scientists call the telescope a Terrestrial Planet Finder.

Artistic renderings of Terrestrial Planet Finder, projected for 2015.
Special space telescopes will search for Earth-like planets around nearby stars. The technology, called interferometry, requires clusters of telescopes in precise formations that relay light waves to a combiner spacecraft.

With planets in view, scientists can examine their spectra. Small, rocky planets that are large enough to hold an atmosphere and close enough to their sun to produce liquid water are the best candidates. The ambient temperature of such planets can be calculated from their spectra. Scientists will search for planets that are not too hot and not too cold, where temperatures are suitable for the origins of life.

If a planet has an atmosphere, that, too, will show up in its spectrum. Plentiful life in complex forms leaves a distinctive signature in its planet's atmosphere. The Earth's atmosphere, with its abundant oxygen and trace amounts of carbon dioxide and methane, could not be maintained in the absence of life. Plants and animals produce atmosphere in such a manner as to create a more favorable environment for their own evolution. Plants absorb carbon dioxide and produce free oxygen; animals produce carbon dioxide and methane. Oxygen interacting with electricity and ultraviolet rays yields ozone, which in turn protects living organisms from harmful light. Earth's atmosphere is a creation and condition of life. The atmosphere that Earth possesses today is not the same atmosphere with which the primitive world began.

Scientists will look for planets with free oxygen and methane in their spectra. In a similar manner, they will seek evidence of water vapor in the atmosphere, which will suggest rain and clouds. The correct amounts of water

"Some scientists working on the question of extraterrestrial intelligence, myself among them, have attempted to estimate the number of advanced technical civilizations in the Milky Way galaxy—that is, societies capable of radio astronomy. Such estimates are little better than guesses. They require assigning numerical values to quantities such as the numbers and ages of stars, which we know well; the abundance of planetary systems and the likelihood of the origin of life within them, which we know less well; and the probability of the evolution of intelligent life and the lifetime of technical civilizations, about which we know very little indeed. When we do the arithmetic, the number that my colleagues and I come up with is around a million technical civilizations in our Galaxy alone. That is a breathtakingly large number, and it is exhilarating to imagine the diversity, lifestyles, and commerce of those million worlds."

—Carl Sagan, astrophysicist and astronomer, 1979.

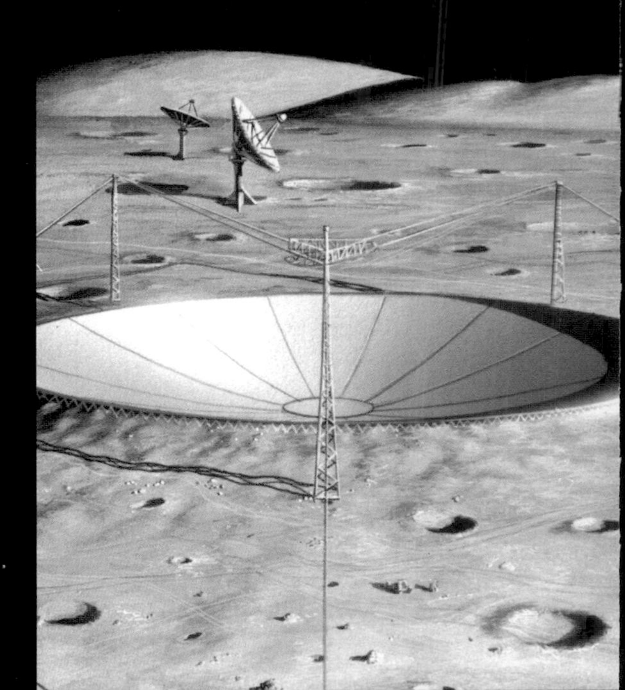

vapor, ozone, methane, free oxygen, and carbon dioxide in the spectra of an extrasolar planet will tell scientists that they have found a place where complex life-forms abound.

Scientists and the public in general will look closely at those spheres. Within fifty years, astronomers will be able to construct space telescopes capable of capturing images of planets orbiting nearby stars with the clarity of Earth photographed from the Moon. The technique for doing this is called interferometry—the same technique used to nullify the glare from distant stars. By cleverly combining the light gathered from many small telescopes, scientists can create sharp images of objects very far away. The first such mission might consist of five Terrestrial Planet Finders, each composed of four telescopes and a relay device. Each set of telescopes collects light and sends it through a relayer to a combiner spacecraft. The combined instrumentation, a Planet Imager, produces fuzzy images that are tantalizingly vague. To produce Earth-like photographs will require increasingly larger arrays, perhaps one hundred planet finders in all. It sounds like science fiction, but it is within human grasp.

No one knows how many habitable planets exist in our stellar neighborhood. No one knows the frequency with which complex life-forms evolve. Some scientists believe that complex life exists everywhere; others believe that it is remarkably rare.

opposite and below

Artistic renderings of radio telescopes on the Moon, after 2025.

Scientists began searching the heavens for radio signals from extraterrestrial civilizations in 1960. To expedite the search, astronomers could construct radio telescopes in craters on the far side of the Moon, where they are shielded from interfering signals from Earth.

Artist's rendering of extra-solar planet, 2000.
A gas giant was discovered orbiting the Sun-like 79-Ceti, located 117 light-years away from Earth in the constellation Cetus. Satellites around such planets may possess conditions conducive to life. Scientists using special space telescopes will capture images of such bodies during the first part of the twenty-first century.

Searching for Intelligent Aliens Humans know from their own experience that complex life-forms build machines. Alien creatures on other planets may be quite noisy, broadcasting signals into the void just as humans have been doing with radio and television waves for one hundred years. In 1960 Frank Drake went to West Virginia and tuned the 80-foot-wide dish of the National Radio Astronomy Observatory to 1,420 megahertz. He pointed it at Epsilon Eridani, a sunlike star some ten light-years away. To his astonishment, he received a strong signal of intelligent origin. Upon further investigation, he was discouraged to learn that he inadvertently had tuned into a secret Earth-based military broadcast.

Since then, a small band of astronomers have been scanning the heavens in search of the proverbial needle in a haystack. During the 1970s, scientists won government funding for the endeavor. They called it SETI: the Search for Extraterrestrial Intelligence. Scientists and their NASA allies asked Congress to accelerate the search by funding Project Cyclops, a $20 billion network of fifteen hundred ground-based antennas. Congress refused and in 1993 terminated all tax-supported funding. To the lawmakers, the project seemed too fantastic to warrant government support. SETI advocates sought and received private funding. Their work continues through the SETI Institute, a nongovernmental body.

SETI scientists use telescopes to collect radio waves from nearby stars. For each observation of a single star, signal detection computers examine tens of millions of channels within a 10-megahertz bandwidth. Computers filter out signals from ground-based sources on Earth. They eliminate signals from the growing number of telecommunication satellites and compare the remaining results with the universe's natural background noise. Any strange or unfamiliar signal triggers a special procedure, called "FUDD," or follow-up detection device. Two separate radio telescopes with their accompanying computers, hundreds of miles apart, track the signal and apply separate FUDDs. It is a tedious process. A signal from an alien civilization could come tomorrow; one may never come at all.

The search for extraterrestrial life drives civil space exploration in the twenty-first century with as much force as the Moon race motivated it during the early years. By the midpoint of the twenty-first century, scientists should have accumulated enough evidence to determine the frequency with which life begins. They may even have discovered places where it has evolved in complex forms.

03 : Lure of the Red Planet

preceding pages

The Martian landscape.

Amateur astronomer Percival Lowell, who thought he saw canals on Mars, inspired Chesley Bonestell to paint water flowing from snowdrifts on the polar cap in 1949 (left). In 1997, the Mars Pathfinder probe landed on the Ares Vallis floodplain. Semirounded stones, tilted in the direction of the flow, suggest that water once flowed in this area of the planet (right).

Mars beckons anyone who has ever looked toward the heavens and wondered how humans came to live on Earth. Perhaps the laws of nature favor the creation of complex life-forms like ourselves. If so, the universe should be teeming with life. Alternatively, circumstances on Earth may be so rare as to make our planet unique in as much of the universe as we shall ever hope to know. After centuries of scientific inquiry, we simply do not know. So far, humans have tried to answer this question by extrapolating knowledge from a single planet: our own. Understanding nature using a single sample—what statisticians call an "n of 1"—is remarkably difficult. It is like trying to understand the whole universe by studying a single sun.

Over the next fifty years, humans will study Mars in exacting detail. The secrets it yields will expand the understanding of the evolution of life—on Earth and elsewhere—in ways barely comprehensible today.

Why Mars? Some experts believe that the processes producing Earth-like planets are exceedingly rare. According to this theory, a habitable planet with a complex variety of surface life requires physical conditions that seldom occur.

The planet must have acquired an orbit that places it in an appropriate zone. It cannot be so close to its star that water boils away nor so far away that it freezes. The planet must be large enough to hold an atmosphere but not so large that its gravity crushes living beings. Astronomers have discovered solar systems that vary considerably from this model. Some planets travel in the right

Artist's concept of a Martian expedition, 1956.

For decades, humans have pondered what they would find by exploring Mars. In this famous Chesley Bonestell rendering, members of the first expedition inspect Martian geology. Technicians have already removed the wings from their landing craft and set the return vehicle upright in preparation for the homeward voyage.

zone but are too large. Others possess elliptical orbits that carry them out of the habitable zone.

The planet needs an atmosphere–not just for surface dwellers to breathe, but to regulate temperatures on the ground. The average temperature on Earth's Moon is well below freezing. Insulating gases like carbon dioxide raise the temperature of Earth about 50 degrees Fahrenheit relative to the Moon. Without that cloaking atmosphere, the temperate climate of Earth would disappear.

Complex life-forms may require plate tectonics for evolution to proceed. Earth's continents float on top of a molten core. As plates collide, landforms rise. Without a moving crust, the continents on which surface dwellers reside might not rise above the level of the sea. More important, continental drift helps to recycle carbon dioxide. In the process of forming shells, sea animals remove carbon dioxide from the atmosphere and deposit limestone. Plate tectonics pulls the limestone toward the molten core, where the carbonate decomposes and reenters the atmosphere through volcanoes. Without this process, carbon dioxide might disappear. Plate tectonics thus acts as a global thermostat helping to maintain surface temperatures in a range conducive to the persistence of life.

Complex life may require a large moon. A large moon creates the tides through which aquatic creatures learn to live in transitional zones. The Moon also stabilizes Earth's tilt. Without that stability, the poles would wobble, causing climatic devastation. Earth's Moon is the result of a freakish occurrence–a cosmic collision that tore off an enormous portion of Earth. None of the other inner planets in this solar system possess a large moon.

"Mars is waiting for us, a big juicy planet with the possibility of water and the possibility of life. Out there looking beautiful and waiting for the Humans to come visit. I think that it is a little strange that the colonization of Mars is such an obvious next step in the exploration of our universe and we aren't recognizing it."

—Hank Green, Mars exploration advocate, 1998.

Mars possesses some but not all of these characteristics. If life began on Mars, something terrible happened to it. Mars lost its surface water; most of its atmosphere disappeared. For people who believe that complex life thrives throughout the universe, Mars presents an enigma. If complex life abounds, something strange occurred on Mars.

Views of Mars. In the early 1960s, Earth-based telescopes revealed large patches that astronomers thought might represent vegetation (below left). They reappear in a 1997 photograph taken through the Hubble Space Telescope (below right).

The Martian Myth One hundred years ago, an amateur astronomer named Percival Lowell offered an explanation for the fragility of Mars. It captured the imagination of people throughout the world. Lowell was the offspring of a prosperous Boston family. Employing the family fortune, he constructed an observatory on a high mesa near Flagstaff, Arizona, from which he began to observe Mars. What he saw shocked the public. Lowell saw canals that appeared to link

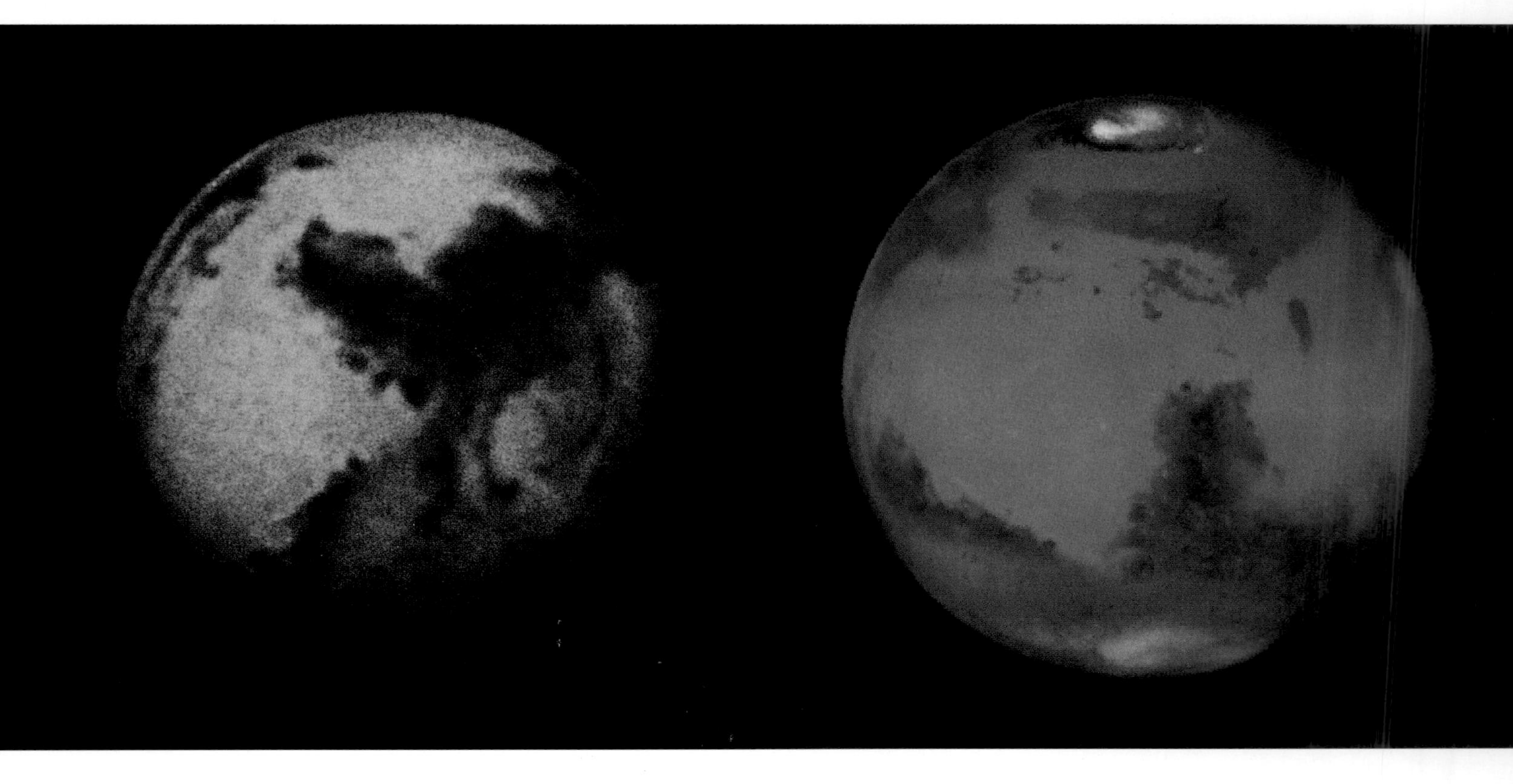

polar ice caps to regions near the Martian equator. In all, Lowell mapped more than two hundred canals.

To Lowell, the canals provided evidence for the proposition that complex beings had evolved on a planet not entirely hospitable to life. Lowell suggested that intelligent creatures had created artificial canals as a means of transferring increasingly scarce water from polar regions to warmer zones. This ancient civilization must be intelligent, Lowell speculated, since only through global cooperation could such a planetary network be maintained. As a prerequisite to global cooperation, he deduced, the Martians must have abolished war. The presence of waterways also suggested that Martians were struggling to survive on an increasingly hostile world.

Lowell's observations gave rise to the Martian myth, one of the most powerful ideas motivating human curiosity about the solar system during the twentieth century. People genuinely expected scientists to find life on Mars. Magazine illustrations commonly portrayed the planet as a network of canals. Editors at *Life* magazine informed readers in a 1944 issue that the canals served to irrigate patches of vegetation "that change from green to brown in seasonal cycles." Willy Ley, one of the most popular science writers of that time, assured readers of a 1952 issue of *Collier's* magazine that plant life "like lichens and algae" existed on Mars. Where plants thrived, Ley added, animals must have followed.

Views of Mars. A popular assumption that canals connected areas of vegetation influenced a 1964 artist's rendering of Mariner IV approaching Mars (below left). NASA spacecraft failed to locate canals but did photograph deep canyons, volcanoes, and local weather, revealed in this mosaic assembled from Mars Global Surveyor imagery taken in 1999 (above right).

Walt Disney, in a widely viewed 1957 television broadcast, showed animated drawings of flying saucers skimming over fields of Martian plants and animal life.

Shortly after Lowell publicized his ideas, the British writer H. G. Wells published a novel further popularizing the idea. In *War of the Worlds*, Wells contemplated how leaders of an ancient but dying civilization would view a more hospitable planet like Earth. They would invade it, Wells suggested. With their superior technology, Martians could snuff out intelligent life on Earth as easily as humans had destroyed primitive cultures on their own planet.

"Mars is the next frontier, what the Wild West was, what America was 500 years ago. It's time to strike out anew. . . . Mars is where the action is going to be in the next thousand years. . . . The characteristic of human nature, and perhaps our simian branch of the family, is curiosity and exploration. When we stop doing that, we won't be humans anymore. I've seen far more in my lifetime than I ever dreamed. Many of our problems on Earth can only be solved by space technology. . . . The next step is in space. It's inevitable."

—Arthur C. Clarke, author, 1992.

Niña, 1492

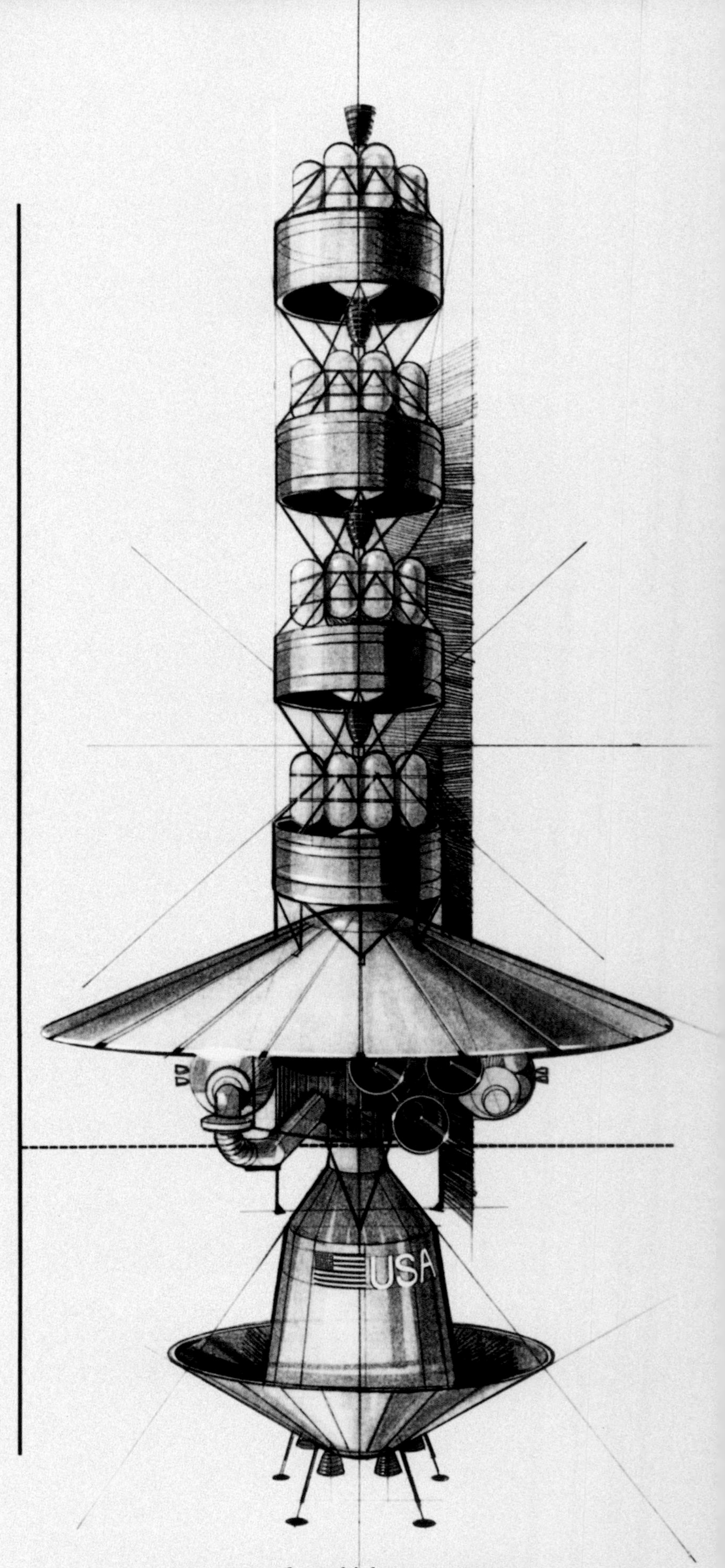

Mars Transfer Vehicle, 21st century

Scarcely a decade later, Edgar Rice Burroughs thrilled a generation of young readers with a series of novels describing the Martian world, or Barsoom, as local residents called it. On Halloween eve, 1938, Orson Welles scared Americans with a radio broadcast reenacting *War of the Worlds*. A movie version appeared in 1953.

"We want Mars to be like the Earth," admitted planetary scientist Bruce Murray. In 1965, humans got their first close-up view. NASA's Mariner 4 sped by the planet, flashing twenty-two photographs back to Earth. All Martian images up to that point had been taken through Earth-based observatories, which produced fuzzy images and optical abnormalities. The close-ups horrified assembled scientists. Nowhere could they see evidence of Lowell's canals. The pictures showed a landscape pocked by craters, much like those scattered across the Moon.

Two Viking spacecraft arrived in 1976. Cameras on the Viking landers took photographs of the Martian landscape; instruments scratched for signs of life in the soil. They found nothing. No living creatures strode across the surface of Mars. The planet seemed no more interesting than the Moon. Science reporter Kathy Sawyer, in one famous comment, observed that Mars was "deader than Elvis."

No large patches of vegetation or artificial canals exist on Mars. Through an inadequate telescope, Lowell mistook surface features like soil coloration and volcanoes for dark spots of vegetation. Perhaps he saw the huge Valles Marineris–the Grand Canyon of Mars–and interpreted it to be a canal. Perhaps in his desire to discover extraterrestrial life, he simply saw large craters and impulsively drew lines between the dots.

Although initial observations discouraged people who hoped to find life on Mars, subsequent findings rekindled hope. Mars is a cold and desolate planet, but this was not always so. It possesses features that resemble coastal plains and river valleys. It is marked by outflow channels that look like the scablands of Washington state, produced by catastrophic floods released by the collapse of natural ice-age dams. Orbiting spacecraft revealed such features in fascinating detail.

At various stages in the history of Mars, huge amounts of water flowed over its surface. In the beginning, Mars may have been covered with lakes and oceans. Rain and snow may have fallen from an atmosphere thicker than the one present today. Substantial deposits of icy soil may exist beneath the surface of Mars. Heated by geothermal activity, groundwater may have percolated up and flowed down crater walls. High-resolution pictures of the Martian surface reveal gullies recently formed on cliffs and crater walls that have all the appearance of being caused by running water.

opposite

Artist's drawing of Mars expedition craft, 1989.
When President George Bush proposed that the United States send an expedition to Mars by 2019, NASA engineers designed an interplanetary spaceship whose size they compared with one of the ships Christopher Columbus sailed.

Photograph of remnants of possible water on Mars, 1997.
NASA's Mars Global Surveyor captured an image of depressions on a crater wall characteristic of water seepage and a dark surface that some interpret as the remains of a lake.

The atmosphere on Mars today is too thin, and the planet consequently too cold, for liquid water to remain on the surface. Due to low atmospheric pressure, any water reaching the Martian surface would boil off in an explosive fashion. Some scientists believe that Mars was always cold and that the waterways appeared during short heating episodes. Others believe that Mars was once warmer and wetter, with a thicker atmosphere.

If liquid water persisted on Mars, single-celled organisms similar to those that appeared on Earth 4 billion years ago should have developed there too. How far such life progressed is as yet unknown. Evidence from the Martian meteorite discovered in the Allen Hills region of Antarctica in 1984 suggests that the process got a start, although scientists do not accept that interpretation as conclusive. If the process began on Mars, at some point it diverged sharply from developments on Earth. Understanding what happened on Mars will help humans to understand the conditions favoring the evolution of life everywhere.

Pursuing the Dream Exploring Mars is a significant challenge. In the twentieth century, NASA scientists succeeded in placing only three robotic spacecraft on the surface of Mars. A NASA polar lander sent in 1999 crashed on entry. Soviet scientists sent seven landers to Mars and failed each time. Experience with orbiters and flybys has not been much better. Of the twenty-nine orbiters and flybys sent to Mars during the twentieth century by scientists from the United States, the Soviet Union, and Japan, 62 percent failed. Mars, it seems, eats spacecraft.

A human expedition to Mars would be very difficult. The longest expedition to the Moon took twelve and one-half days. A Martian expedition would be gone three years. The crew would face radiation hazards, equipment failures, the risks of landing, and the dangers of liftoff. The mobilization back on Earth necessary to mount such an expedition would be more complicated than any engineering endeavor ever attempted and could potentially cost hundreds of billions of dollars. Technically, NASA engineers could mount such an expedition by 2030. Political and financial obstacles to such an endeavor may prove insurmountable, given the costs and obvious dangers. As discouraging as it seems, humans will probably choose to delay the long-held goal of a Martian expedition, concentrating instead on robotic exploration for the first half of the twenty-first century.

Such facts have not discouraged space advocates from trying. In 1989 President George Bush proposed that the United States organize a human expedition to Mars. He set, as a target date, the fiftieth anniversary of the first landing on the Moon: 2019. NASA engineers planned a colossal undertaking incorporating space stations, interplanetary spacecraft, gigantic rockets, lunar bases, and a Martian research station.

Artist's depiction of a geologist exploring Mars, after 2030.
Hanging more than 2 miles above the Tharsis Bulge, a geologist-astronaut collects samples from the hardened lava cliff face that rims Olympus Mons, the solar system's largest known shield volcano. Mars offers a laboratory for studying the processes required to support life on Earth.

The proposed Mars spacecraft was as massive as a football field. It contained a cluster of engines adequate to push the spacecraft toward Mars and a big aeroshield capable of braking the spacecraft once it got there. The spacecraft also held a transit module for the interplanetary voyage and a crew excursion vehicle for landing on Mars. The excursion vehicle carried its own aeroshield for further braking and descent, along with all of the fuel it needed to land and take off from Mars.

The spaceship was so big that it could not be launched from Earth. Instead, NASA engineers planned to assemble it at an orbital drydock, a new feature of the International Space Station, then mate it with engine clusters a safe distance away. To deliver the necessary equipment into orbit, NASA officials proposed a rocket larger than the Saturn V that lifted Americans to the Moon. Officials priced the spacecraft, the rocketry, and the necessary orbital infrastructure, threw in a lunar base, and presented a cost estimate totaling $450 billion. That sum equaled one-half of all federal tax revenues in the year that the initiative was proposed.

The whole package was an engineer's dream. It required the development of every new technology in NASA's long-range plan. The spaceship was especially ambitious. Robert Zubrin, an aerospace engineer and Mars advocate, called it a "death star." The magnitude of the proposal encouraged lawmakers to kill the initiative before NASA could start spending money on it. Members of Congress refused to approve even the modest budget requests needed to start thinking about the technology necessary for the voyage.

Zubrin insisted that NASA planners rethink their mission plans. Specifically, he suggested that the first humans to Mars "live off the land." If Meriwether Lewis and William Clark had been forced to carry all their fuel and supplies across the continent, they would never have reached the Oregon coast. They would have been stuck forever in Missouri. In a similar fashion, Zubrin proposed that the first humans on Mars extract fuel and consumables from the Martian environment, and he built a machine to show how it could be done.

In 1992 Daniel S. Goldin became NASA administrator. Confronted with the demise of the Mars expedition—the so-called Space Exploration Initiative—Goldin challenged NASA employees to develop "faster, better, cheaper" methods of spaceflight. NASA engineers began with small robotic spacecraft, including the highly successful Mars Pathfinder mission that landed on Mars in 1997. Workers at the Jet Propulsion Laboratory constructed and flew the Pathfinder spacecraft for one-fourteenth the cost of a traditional planetary mission. Simultaneously, Goldin encouraged engineers at NASA's Johnson Space Center to start thinking about a human expedition with similar cost savings.

A Low-Cost Mission to Mars Advocates of the "faster, better, cheaper" approach believe that a human expedition to Mars could be conducted for about $50 billion. In inflation-adjusted dollars, that is just one-third of the total cost of sending humans to the Moon. Such an undertaking would require a Mars exploration program funded at a level around $6 billion per year over eight years, a diminutive sum given the size of the United States economy. The first landing would occur within nine years of the decision to go, and landings would continue every two years thereafter.

Skeptics say that this is impossible and point to the failure of three "faster, better, cheaper" Mars probes in 1999. Those were robotic spacecraft—an orbiter, a polar lander, and a pair of microprobes. Landing humans and keeping them alive are harder than flying robots. Mission cost, the skeptics insist, quickly rises with mission complexity.

Who is right? Without trying, it is hard to tell. Mars advocates want to try. If they can convince enough Americans to support the expedition, and a sufficient number of politicians to fund it, the mission as planned would occur in the following manner.

Six years after funding begins, NASA engineers transport three large rockets to the Kennedy Space Center. In some respects, each rocket resembles a NASA space shuttle—two large boosters attached to a 27.5-foot-wide external fuel tank supplemented by hydrogen-burning engines on the aft end. In place of the familiar orbiter spacecraft, the rockets support a tall cylindrical-shaped object. Each cylinder holds 176,000 pounds of payload, bound for Mars.

The first cylinder contains a habitat module, a rocket motor, and consumables for the first crew. It goes into orbit around Mars, where it waits for the homebound crew. This is the transfer vehicle that will bring the first expedition back from Mars. The second cylinder contains a habitat module outfitted as a scientific laboratory, a nuclear power plant for generating electricity on the surface of Mars, an unpressurized rover, and more consumables. It lands on the surface of Mars. The third cylinder delivers more rovers (pressurized, unpressurized, and robotic) and the most critical pieces of equipment for the voyage home: an ascent vehicle for leaving Mars and a manufacturing plant capable of producing its own propellant. The third cylinder also lands on Mars.

Fuel is manufactured on Mars from the local atmosphere, consisting mainly of carbon dioxide. Martian air is pumped into a

reaction chamber in the manufacturing plant, mixed with liquid hydrogen, and heated. The resulting process, discovered in the nineteenth century by French chemist Paul Sabatier, produces methane and water. The methane is pumped through a cryogenic cooler, which reduces it to a liquid state, and is stored for use as rocket fuel. The resulting water is pumped into an electrolysis unit where electrodes separate it into hydrogen and oxygen. The hydrogen refuels the Sabatier reaction; the oxygen is cooled and stored for future use as oxidizer for the rocket fuel and breathing air.

The total process produces rocket fuel, oxidizer, water, and surplus oxygen using one small manufacturing plant and some liquid hydrogen brought from Earth.

The expeditionary team does not arrive with the first three spacecraft. Appropriately, the astronauts wait until the manufacturing plant has produced sufficient fuel. At the next conjunction of Earth and Mars, twenty-six months after the first launch sequence, three more rockets appear. Once again, they contain a transfer vehicle bound for Martian orbit, another manufacturing planet, and another habitat module bound for the landing zone. This time, however, the landing habitat carries a crew–the first humans to visit Mars.

In Earth orbit, the crew checks out their spacecraft. They depart for Mars, using a relatively fast transit of 150 days. This requires significant improvements in propulsion technology. To achieve the necessary thrust, engineers propose a nuclear thermal propulsion engine. The engine employs a very hot nuclear core to increase the temperature of liquid hydrogen. The rapidly expanding hydrogen gas provides three times as much thrust as conventional chemical fuels.

Artist's depiction of a low-cost Mars expedition, about 2020.
Rockets launch a prototype for a Mars habitat (opposite top), which is tested on the International Space Station (opposite center). An automated factory is sent to Mars to make propellant for the return voyage (opposite bottom and above top). The first crew arrives in a new habitat module carrying a pressurized rover (above).

As it approaches Mars, the spacecraft must slow down. This is accomplished through aerobraking. The spacecraft dips into the Martian atmosphere, where friction builds up against a cone-shaped aeroshell surrounding the spacecraft. Two aerobraking maneuvers are required–one to place the spacecraft in Martian orbit and a second to begin the landing phase.

The crew uses three mechanisms to land. The aeroshield slows the spacecraft in the upper atmosphere. Parachutes further slow the spacecraft once it dips into the thin Martian atmosphere. Rocket engines burning methane and liquid oxygen place the spacecraft on the ground.

Upon arrival, the expeditionary force spends six hundred days exploring Mars. At first, they concentrate on establishing their base camp. They deploy an inflatable greenhouse in which they grow food and observe the behavior of plants brought from Earth, now growing on a world with one-third the

gravity of their home planet. Once the Martian outpost is secured, the crew dispatches automated rovers into the surrounding territory. Based on information gathered from robotic forays, the crew begins expeditions of their own. They venture 60 miles from their home base, exploring an area about the size of Switzerland.

The crew collects rock samples for analysis in a small laboratory set up in the habitat module. Using samples from the Martian landscape, they replicate the experiments performed on the Allen Hills meteorite. Outside the spacecraft, they drill into the Martian substrata in search of ice water and subterranean life. They venture to nearby canyons and chasms where they observe the geological history of Mars. They search for fossils. They locate natural resources whose presence has been detected by satellites orbiting Mars. They retreat to the security of their home base during solar flares.

As their departure date nears, crew members practice launch procedures and check equipment. Strapping themselves into their ascent vehicle, they leave the surface of Mars. They rendezvous with the orbiting habitat module, launched from Earth years earlier. Transferring to the habitat module, they hope the engines will start. They return home. The trip takes 110 days. The first humans to visit Mars land on Earth "Apollo-style"—directly reentering the atmosphere for a parachute-assisted landing.

Back on Earth, a replacement crew departs for Mars along with another set of vehicles. When the second crew arrives, they find three manufacturing plants and three habitat modules at the landing zone. Two return vehicles circle Mars, along with an infrastructure of orbiting satellites and ground stations. Humans have established the first elements of a research station on the surface of Mars, with more people arriving every two years.

The technical problems confronting mission planners are considerable. The crew is exposed to two types of radiation: cosmic radiation produced by the galaxy as a whole and solar flares emitted by the sun. A fast transit time is the best protection against galactic radiation; the local atmosphere protects the crew once on Mars. Solar flares can be lethal, especially in the unprotected vacuum of space. Engineers can shield the crew with water, creating a donut-shaped water tank with living quarters in the middle into which the crew retreats until solar storms subside.

Procedures to compensate for the absence of gravity during the interplanetary voyage are necessary to maintain a healthy crew.

Engineers contemplate use of a tether stretching between the spaceship and its empty propulsion stage. By rotating the two vehicles, astronauts can create artificial gravity. The procedure adds weight, however, and substantially complicates flight operations. As an alternative, engineers consider a rigorous exercise schedule.

For the first phase of the expedition to succeed, engineers must place four spacecraft on the surface of Mars. Engineers struggle to build a spacecraft sturdy enough to survive multiyear voyages and land on an alien planet, and to do so at cost levels promised by mission planners.

Cost is a major obstacle. The temptation to spend too much money is strong, especially as technical risks become well known. Environmentalists object to the use of nuclear propulsion. The nuclear engine must be launched from the Kennedy Space Center, even though it is not fired until it reaches space.

If humans decide to go to Mars in the next fifty years, it will not be because the mission is easy or technically feasible. Ultimately, it will happen because people want to go so much that they overpower the inevitable obstacles.

The most complete Mars exploration effort requires robots and humans working together: Robotic satellites in Martian orbit provide reliable communication channels, map the Martian surface, and track storms. Automated rovers precede humans into unknown Martian terrain. A solar-powered model airplane without a human pilot onboard flies through the local atmosphere. A similar mission could be conducted with balloons. Humans then follow to the most interesting sites.

Mars may turn out to be a lifeless world, with wet periods too brief to allow life to begin. If Mars is as lifeless as the Moon, interest in exploration will decline.

More likely, Mars once possessed conditions similar to those found on Earth. Such a planet provides a fantastic laboratory for learning more about the ways in which habitable planets evolve. Exploration will continue for many decades as humans probe the processes that create habitable worlds.

Artist's depiction of a low-cost Mars expedition, about 2020. The crew uses a pressurized rover to explore Mars (opposite top). After spending six hundred days on Mars, the crew departs in the ascent vehicle set on top of the automated fuel factory (opposite center), docks with another habitat module orbiting Mars (opposite bottom), and returns to Earth (above).

following pages
Artist's concept of the first humans on Mars, about 2020. In this depiction by Ren Wicks, astronauts record shifting wind speeds in anticipation of an approaching dust storm. Mars advocates would like to dispatch the first expedition by 2020, but the cost and technical complexity will undoubtedly delay it.

We still possess the greatest gift of the inheritance of a 400-year long Renaissance: To wit, the capacity to initiate another by opening the Martian frontier. If we fail to do so, our culture will not have that capacity long. Mars is harsh. Its settlers will need not only technology, but the scientific outlook, creativity and free-thinking individualistic inventiveness that stand behind it. Mars will not allow itself to be settled by people from a static society—those people won't have what it takes. We still do. Mars today waits for the children of the old frontier, but Mars will not wait forever."

—Robert Zubrin, author and Mars expedition advocate, 1996.

04 : Getting There

* * * * * *

preceding pages

Evolution of technology. Robert H. Goddard's experiments with liquid-fuel rockets led to the development of ballistic missiles, orbiting satellites, and interplanetary spacecraft. He launched the first rocket from a field near Worcester, Massachusetts, on March 16, 1926 (left). Although it rose only 41 feet, it proved the basic technology for the machines that followed. Seventy-one years later, NASA sent the Cassini space probe on the first step in its long journey toward Saturn, using a Titan IV (right). With a total stack height equivalent to that of a twenty-story building, the rocket is the largest expendable launch vehicle in the U.S. fleet.

Access—no single word better describes the primary concern of people interested in space. All participants–civil, military, and commercial–need affordable, reliable, frequent, flexible access to space. Too often they do not get it.

Shipping companies on Earth deliver goods on a fairly predictable schedule. They operate in a variety of climates, book shipments on short notice, and rarely destroy their cargo. The same cannot be said for space. Technical glitches and adverse weather frequently delay delivery. Flights need to be booked two to four years in advance, and the chances that cargo on a large launcher will be destroyed range from 2 percent to an embarrassingly high 18 percent.

Pioneers of spaceflight believed that humans could make space travel safe and inexpensive. They gained confidence by watching aeronautical engineers develop jetliners that moved people through the air. Despite years of effort, however, the dream of cheap and easy space access has not been attained. Costs remain particularly high.

During the next fifty years, rocket scientists will try again to achieve that dream. They will do so first through a series of changes to chemical rockets. From the first experiments by Robert H. Goddard in the 1920s through the mighty Saturn V Moon rocket, engineers have relied upon chemical rockets to achieve their goals. Chemical rockets, however, are notoriously inefficient and costly to operate. They will be followed in the twenty-first century by alternative forms of propulsion, some drawn from the pages of science fiction.

Rocket Basics Experiments with chemical rockets in the twentieth century allowed humans to reach space for the first time. The rocket is a reaction device, based on Sir Isaac Newton's Third Law of Motion: "For every action there is an equal and opposite reaction." The pressure of burning gas creates the action of a rocket. As gas escapes through a nozzle at ever-higher velocities, it creates thrust. The remaining gas presses against the forward wall of the engine and propels the rocket upward. The rocket continues to accelerate until all of its propellant is gone.

Twentieth-century rocket pioneers systematically applied this basic principle to engines of increasing complexity. Technological understanding allowed Robert Goddard to launch the first crude liquid-fuel rocket in 1926. Further developments led to the German V-2 of World War II, to ballistic missiles with intercontinental range, and finally to the most complex launch system ever devised: NASA's space shuttle. Astronauts flew the shuttle into space for the first time in 1981; the vehicle will remain a workhorse for human flight initiatives well into the twenty-first century.

Photograph of preparations for a V-2 launch, 1946.
The godfather of virtually all chemically fueled rockets is the V-2 ballistic missile, developed by German rocket engineer Wernher von Braun and his Pennemünde research team during World War II. At war's end, American military officers brought members of the von Braun team and their V-2 rockets to the United States, where they formed the basis of the nascent American rocket development effort. In this photograph, American technicians prepare to launch one of the captured V-2s at the White Sands Proving Grounds, New Mexico.

The path of rocketry from Goddard to the space shuttle rests on one simple mathematical equation. The product of the propellant mass flow (m) and its exhaust velocity (C) equals the thrust (T) generated by a rocket: $T = mC$. The higher the exhaust velocity, the more thrust generated per propellant mass flow. Thrust gives the rocket the kick necessary to overcome gravity and reach orbit. It does the heavy lifting necessary for spaceflight and in that sense is analogous to the horsepower rating used in automobiles

Most launch vehicles are characterized by the thrust they generate and the payload mass they deliverer to orbit. Some payloads go to low Earth orbit (LEO), 100 to 350 nautical miles high, where they zip around the planet at a speed of 17,500 miles per hour. Others travel to geosynchronous orbit (GEO), where at 22,300 nautical miles their speed matches the rotation of Earth, placing them at a stationary point in the sky. Each of the space shuttle's three main engines generates an average thrust of 470,000 pounds. With the help of two solid rocket boosters, which do most of the lifting, the engines deliver 53,000 pounds of payload to LEO. The shuttle cannot travel to geosynchronous orbit, but it has carried rockets capable of pushing 13,000-pound objects to GEO. Payload delivery capabilities rise or fall somewhat depending on exact altitude

and on the inclination of the orbit over the Earth's surface. Many aerospace engineers make a living calculating to multiple decimal points these critical parameters for every space launch.

The second measurement used by rocketeers to rate engines is known as specific impulse (Isp). This represents the number of seconds during which a rocket engine can produce 1 pound of thrust from 1 pound of propellant. The Isp, in effect, measures the efficiency of the rocket engine. If thrust is analogous to horsepower in an automobile, Isp corresponds to miles per gallon of gas. The best chemical rocket engines, the ones that power the major launch vehicles, are limited to an Isp of about 500 seconds. This represents the current state of the art in launch technology.

Chemical propellants limit Isp. Hydrazine, the oily substance that powers many small rockets, has a value of about 200 seconds. A combination of kerosene and oxygen–the fuel used by von Braun in developing the German V-2–has an Isp of about 350 seconds. By mixing liquid hydrogen and liquid oxygen, shuttle designers can generate an Isp of 450 seconds. Depending on the complexity of the engine, the efficiency of the nozzle, and a host of other attributes, engineers can tinker around these boundaries to deliver a slightly higher Isp for any given chemical engine.

The Isp delivered by chemical rockets is quite inefficient, especially for long-duration interplanetary flights. Because of these limitations, robotic missions to the outer solar system typically fly by other planets on their way. This so-called gravity-assist trajectory allows them to accelerate toward these bodies and skim away from them in the manner of an object being thrown out of a slingshot. Such trajectories limit launch windows to periods when the planets are properly aligned and can take years to complete. Flights through the inner solar system take considerable time as well. Some of the routes for a human mission to Mars using chemical rockets take as long as three hundred days for the transit phase, one-way. Given the potential exposure to radiation, this is the outer limit for human voyages.

Fortunately, other types of propulsion offer much higher specific impulse and will eventually open a new age of planetary exploration. A nuclear rocket engine has the capability of generating 900 seconds. That is small when compared with the 2,000 to 20,000 second Isp generated through electrical or ion propulsion systems. So why not move to these new types of rockets? Because of other key factors in rocket technology: flow rate and engine thrust. Flow rate defines the pace at which the propellant can be supplied.

Ion propulsion engines are limited in their propellant flow rate. They are exceptionally efficient, but low flow rates steal their power. Once beyond the gravitational pull of Earth, ion engines generate a great deal of energy over long

Artistic renderings of large launch vehicles, early 1950s. To build space stations and reach the Moon, engineers believed that they needed large, three-stage rockets capable of propelling humans and cargo into low Earth orbit. In these two conceptions, the winged third stage is reusable, capable of flying back to Earth like an airplane. Rolf Klep's cutaway (right) shows how the stages fit together.

United States

periods of time using extremely small amounts of fuel. With steady application, this speeds long-distance travel beyond anything possible with chemical rockets. Nuclear rockets have the potential to generate both high Isp and high thrust, but the risk of a radioactive accident during launch enjoins their use.

Photograph of space shuttle Discovery, 1988.
Astronauts flew the first reusable launch vehicle into orbit on April 12, 1981. Seven years later, with dawn approaching, technicians prepared Discovery for its return to space—the first flight following the loss of Challenger in 1986.

Launch Costs Space travel started out as and remains an exceptionally costly affair. The best expendable launch vehicles can deliver payloads for a cost of about $5,000 per pound from Earth to low Earth orbit, although many providers charge more. As a result, space travel has remained the province of institutions with access to large sums of money—tax-supported bureaucracies, a few high-end communications satellite companies, and other unique users.

Modest space launchers, delivering relatively small satellites, average $25 to $50 million per flight, or about $10,000 to $40,000 per pound. The Saturn V moon rocket, the most powerful launch system ever developed, produced at launch 7.5 million pounds of thrust. It could place into orbit a payload of 285,000 pounds. It did so, however, at the prohibitive cost in year 2000 dollars of $1.3 billion per launch. That produced a relatively low cost per pound of about $4,500, but gargantuan launch costs. And these are just launch costs—they do not include the price of satellite development or the sums that must be spent on payload preparation, insurance, boosts to higher orbits, ground support, and flight operations. Spaceflight is a rich person's game.

In 1972 NASA promoted the reusable space shuttle as a means of reducing the high cost of flight. Some NASA officials compared the use of expendable launch vehicles like the Saturn V to a railway engineer throwing away the locomotive and boxcars after every trip. In today's dollars, NASA officials sought to cut cost to orbit from $10,000 to $1,000 per pound.

In spite of high hopes, the shuttle provides neither inexpensive nor routine access to space. It launches its 53,000-pound payload at a cost of about $8,500 per pound, or $450 million per flight. Shuttle advocates hoped that a low-cost shuttle would underprice alternative launch systems. This did not occur. The cost per flight is so high that only the United States government can afford to fly it. During its design, NASA officials predicted that the shuttle would fly twenty-four times per year. In fact, shuttle managers took five years to accumulate their first twenty-four flights. Turnaround time between flights requires months of work rather than days. While the shuttle orbiter is reusable, its cost and complexity, coupled with the ever-present rigors of flying into space, severely limit its use.

The space shuttle is both a triumph and a tragedy. NASA operates an exceptionally sophisticated vehicle, one that no other nation on Earth could have built during that period. At the same time, flying the shuttle is essentially a continuation of space spectaculars à la Apollo, continued as much for national prestige as for the efficiencies involved. The shuttle's much-touted capabilities remain unrealized. Shuttle managers have made far fewer flights and conducted far fewer scientific experiments than NASA officials publicly predicted.

Most people who want to deliver payloads into space rely on launchers—Atlas, Delta, and Titan class vehicles—that began their development as intercontinental ballistic missiles during the 1950s. In 2000, government officials in the United States relied on the descendants of these three ballistic missiles for the bulk of their space access requirements. The most recent version of the powerful Atlas rocket, the Atlas IIAS, began flying in 1993 and can send 19,000 pounds into low Earth orbit at the bargain price of $105 million per flight, about $6,000 per pound. The three-stage Delta II entered operational service in 1989 and can place 3,190 to 4,060 pounds into orbit at a cost of $45 to $50 million per launch, about $12,000 per pound. The Titan IV, the most powerful expendable launch vehicle operated by the United States, can place 39,000 pounds in low Earth orbit. It costs more than $240 million per flight, or $6,000 per pound.

Second-generation reusable spacecraft.
Efforts to create less costly, more reliable successors to the NASA space shuttle include the Delta Clipper, shown in its first test launch in a 1996 photograph (far right). Technicians achieved a twenty-six-hour turnaround between the second and third test flights, but the craft suffered damage during the fourth flight. A year later, Congress created the X-33 program, hoping to spur commercial investment in a full-scale VentureStar launcher (shown in an artist's rendering, top) by providing public funds for a smaller experimental vehicle. The X-34 (center) was created to test new technologies at speeds up to Mach 8 and altitudes 50 miles high. NASA hopes to fly redesigned versions of these vehicles by 2010. The X-38 (bottom) is intended to serve as an emergency crew return vehicle for the International Space Station.

Second-Generation Launchers Everyone in the aerospace business agrees that first-generation launch vehicles are too expensive, too unreliable, and too hard to maintain. To serve the needs of twenty-first-century users, second-generation launchers must begin to fly. On paper, second-generation launchers promise admirable improvements relative to old launch vehicles. If rocket scientists have their way, the first models of the second generation will begin routine operations by 2010.

NASA officials developing second-generation launchers have invested heavily in reusable X-vehicles: the X-33, X-34, and X-38. The X-38 may enter service as the emergency crew return vehicle for the International Space Station. Although the X-33 program has been cancelled, NASA still wants to develop low-cost vehicles like these. Developing an operational spaceflight vehicle out of the X-plane mold is not likely to occur, however. Watch instead the large number of private companies attempting to resolve the space access problem with new nonpiloted launchers of their own. Major players in the launch service industry are racing to develop new lifters. A $10 million prize awaits the first private launch service with a rocketship that can loft three people to 62 miles twice in two weeks. This so-called X-Prize has raised corporate interest in space access and pumped up the charisma of the space access business.

New initiatives include updated versions of existing rockets such as the Lockheed Martin Atlas series, the Orbital Sciences Corporation Pegasus XL and Taurus rockets, and the Boeing Company's Delta III. Private entrepreneurs are also in the race: Kelly Space and Technology's Astroliner, Rotary Rocket

United States
Orbital
X-34

Artist's concept of a third-generation test vehicle, projected for about 2025. Rocket scientists hope to reduce the weight of space launchers by eliminating much of the oxidizer they must carry. This Rocket Based Combined Cycle launcher would scoop in oxygen from the air as it speeds toward space.

Company's Roton, the Sea Launch Company, Kistler Aerospace Corporation's K-1, and Beal Aerospace's BA-2. All of these rocket entrepreneurs want to capture a share of the space access market.

NASA officials hope that the development funds spent on the X-33 will inspire Lockheed Martin to raise private capital for a large, single-stage-to-orbit vehicle called VentureStar. To qualify as a second-generation launch vehicle, VentureStar would have to meet a forbidding set of requirements. Engineers would have to reduce risk to the point that the probability of losing a craft is no worse than 1 per 10,000 missions. (The odds of a shuttle catastrophe are 1 in 450. As a point of contrast, military pilots face the probability of a catastrophic loss in every 22,000 flights, and commercial airline passengers take a risk of 1 in 10 million.) VentureStar designers would have to reduce the cost of delivering a pound of payload to low Earth orbit from $10,000 to $1,000 per pound in year 2000 dollars. They would have to provide a reliable escape system for the crew, similar to the ejection seat on a jet fighter, a feature that the current space shuttle does not provide. They would have to build a vehicle that could be prepared for flight in seven days. (Technicians servicing the space shuttle need as long as five months.)

Is the hope for an efficient second-generation launcher wishful thinking, or can it really be done? The answer depends upon the ability of engineers to squeeze more performance out of chemical rockets. Although few people want to depend upon them in the long run, chemical rocket engines will power

second-generation launchers. Engineers will try to improve turbomachinery, combustion devices, lines, and ducts. Turbine-driven centrifugal pumps deliver fuel and oxidizer to rocket engines. Combustion chambers provide thrust and produce hot gas that can power turbines. Lines, ducts, and valves serve as the plumbing system that routes various hot and cold fluids through the engine. High-performance components such as these are hard to manufacture and endure unnatural stress while in operation.

Chemical rockets can be made more efficient through new engines with greater efficiency. Engines under development waste less energy–engine temperatures can be decreased, mass flow rates reduced, and engines made smaller–thus increasing specific impulse. New alloys enable engines to withstand greater heat, endure less erosion, achieve longer life cycles, and require less maintenance. To achieve desired properties, some alloys are produced in the form of powders that are molded into desired shapes under heat and pressure.

Second-generation launchers incorporate a number of improvements to the rocket body. Consider this: the faster a craft flies through the atmosphere, the hotter the leading edges of the fuselage and any wings become. To combat the heat problem, engineers design spacecraft with blunt surfaces presented to the air and cover these surfaces with insulating materials like ceramic tile. Blunt surfaces dissipate heat more effectively, but produce drag, which reduces lift during the launch phase and translates into steeper descent rates during landing.

By improving thermal protection systems with lighter, more heat-resistant materials, engineers can build rocketships with sharper lines. This produces less drag. A winged space shuttle constructed out of lighter materials can fly farther, thereby allowing incoming pilots to choose among a wider selection of landing sites. A multitude of landing sites means that space-based enterprises can locate ground operations wherever needed to support their business plans, or simply that adverse weather can be avoided. Advanced thermal protection is only one example of a key improvement in launch technology.

Composite structures represent another advance. In the last part of the twentieth century, composite structures appeared in many mechanisms, from race cars to aircraft to bicycles. Manufacturers create composite structures by combining different materials such as carbon fibers and epoxy. Composite structures appear where high strength, stiffness, and low weight are important. They promise to make spacecraft lighter, stronger, and more resistant to heat. An attractive application for these honeycomb-shaped materials is in the fabrication of cryogenic fuel and oxidizer tanks.

If any of these new technologies produce rockets that can reduce launch costs by the much-proclaimed factor of ten, then watch out–all kinds of possibilities emerge.

" . . . We have to invest in the future. I believe we have no right, as a society, to say that because we have problems in the present, we will walk away from the future. The civil space program is investing ten years, twenty years, centuries ahead.

NASA's budget is $14 billion. Our federal budget is $1.5 trillion. We could take the whole NASA budget and, in a feeding frenzy in the U.S. Congress, vaporize that budget in two hours. And even if we converted it into valuable things instead of pork, nine-tenths of one percent is not going to solve this nation's fundamental problems. But I would weep for our nation if we didn't have a space program."

—Daniel S. Goldin, NASA administrator, 1992.

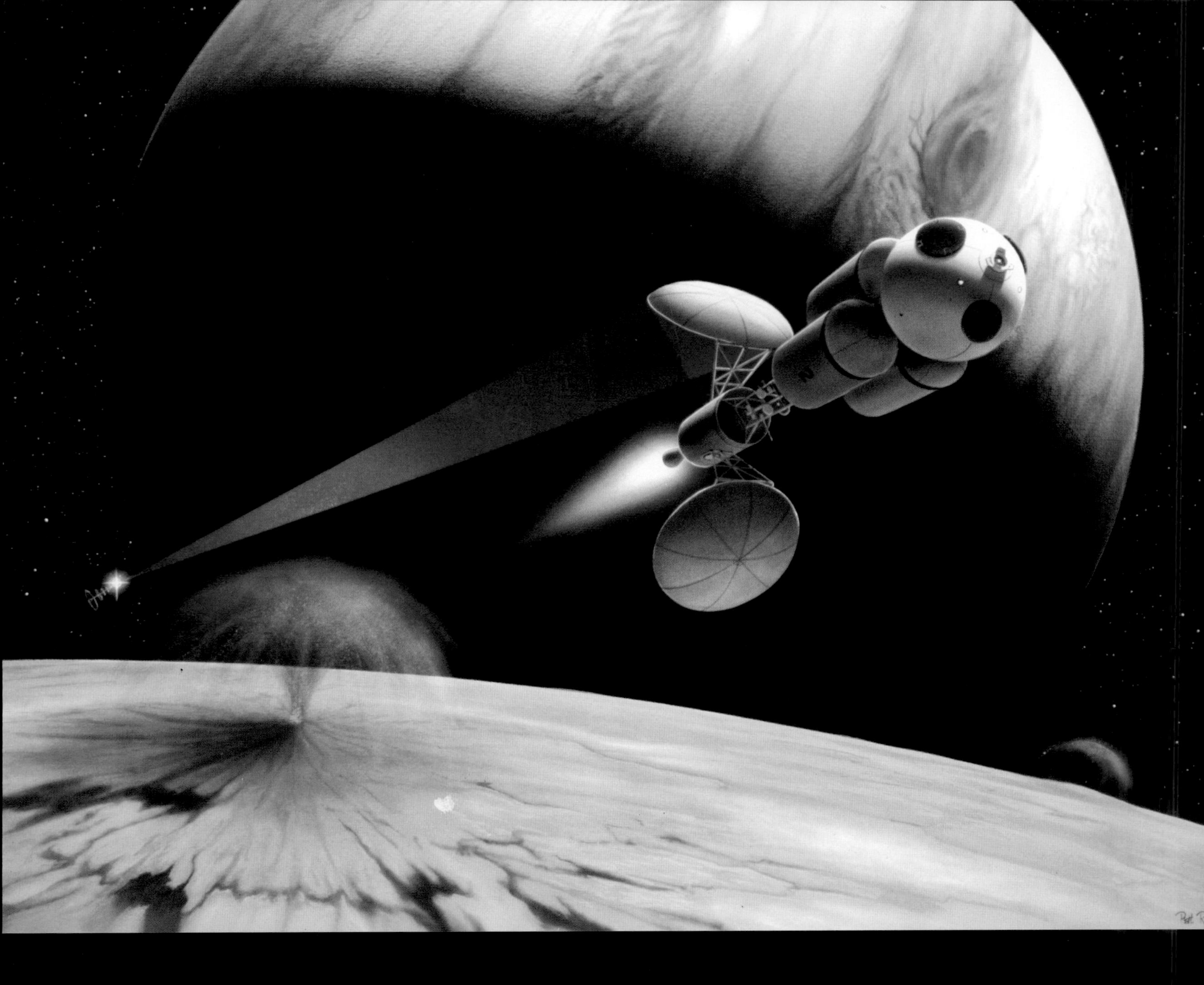

"Too many of us have lost the passion and emotion of the remarkable things we've done in space. Let us not tear up the future, but rather again heed the creative metaphors that render space travel a religious experience. When the blast of a rocket launch slams you against the wall and all the rust is shaken off your body, you will hear the great shout of the universe and the joyful crying of people who have been changed by what they've seen."

—Ray Bradbury, author, 1995.

Making Spaceflight Routine Could third-generation launchers appear before 2050? Absolutely, and they will begin to relieve the most potent restrictions imposed by conventional chemical rocketry. Beyond the space shuttle and the second-generation launchers of 2010, engineers will attempt to produce third-generation rocketships that could come on-line around 2030. The specifications for these vehicles, if attained, could make spaceflight as routine as air travel. They would truly open the space frontier.

NASA planners want to produce third-generation launch vehicles that can improve safety to the point where the probability of losing a craft is no worse than 1 in 1 million, a ratio that approaches the safety level of commercial jetliners. Planners want to reduce payload costs by another factor of ten, to about $100 per pound. They would like to develop vehicles that can be prepared for launch by a handful of people in a day or two, allowing thousands of flights per year. Third-generation reusable launch vehicles that meet these goals will expand space commerce and provide a departure platform for new destinations. Such specifications are easy to state but hard to achieve.

Third-generation launchers will still utilize chemical engines, but technology will improve to the point where a much greater portion of the system can be devoted to payload rather than propulsion. The chief development will be a significant reduction in the need for onboard propellant. For example, instead of carrying all of the oxidizer needed to produce combustion, such as liquid oxygen, engines may be designed to burn air gathered from the atmosphere for an extended period of time.

Throughout the history of chemical-fueled rockets, engineers have achieved "liftoff" by placing rockets on end and firing engines toward the ground. The rocket goes straight up in the air, fighting gravity all the way. NASA's plans for a Spaceliner 100 revises this technique. Spaceliner 100 is not a program directed at the production of a specific vehicle, but rather a technology-development effort seeking to make such vehicles possible. A Spaceliner 100 rocket would be launched horizontally along a track. Using magnetic levitation to eliminate friction, linear electric motors would accelerate the vehicle to more than 1,000 miles per hour before it leaves the track. This MagLifter launch facility will enable space vehicles to be launched toward orbit in a more airplanelike manner, like jetliners gathering speed as they roll down runways. MagLifter catapults will provide launch assistance to a variety of vehicles accelerating down a guideway for 2 to 5 miles. Near the end of the guideway, the track will turn upward at an angle of 55 degrees, whipping the rocket into stage one of its journey into space.

Having achieved speeds in excess of 1,000 miles per hour while

opposite
Artist's concept of a laser-powered rocket, after 2040.
Exotic technologies may power rocketships on future flights. A power station, drawing energy from Jupiter's moon Io, could beam energy through a laser to an electric-powered rocket engine. NASA tested such an engine in 1998.

Artist's depiction of a pulse propulsion system, about 2025.
The most famous spaceship in modern science fiction, the Starship Enterprise, uses anti-matter pulses, called "warp drive," to achieve exponential acceleration. High-velocity bursts can be more efficient than slow constant burns for certain rocket engines.

Artist's concept of MagLifter, after 2030.
Third-generation spacecraft will carry much less onboard propellant than first-generation spacecraft like the space shuttle. This can be accomplished by launching a spaceship horizontally along a track that magnetically levitates and accelerates it to supersonic speeds. Rocket engines take over as the track bends upward and hurls the rocketship toward space.

still on the MagLifter, spacecraft pilots can start their engines in a ramjet mode. Ramjet engines rely upon the injection of fuel into a stream of air compressed by the forward speed of the jetcraft. This reduces the need for liquid oxidizers or turbofan compressors.

Engineers hope that horizontal takeoff techniques will simplify ground operations–maintenance, cargo loading, and passenger boarding. Through further simplicity, they hope to enhance reusability, cut costs, and produce faster turnaround times.

With their air-breathing engines and sharp wings, third-generation launchers will resemble ultramodern jetliners. In fact, the distinguishing line between rockets headed for space and aircraft flying through the atmosphere may disappear. Advanced aerospace vehicles will be capable of operating in a seamless regime from Earth's surface all the way to orbit, as appropriate to each flight's intended destination.

Rockets Beyond Before the midpoint of the twenty-first century, the first prototypes for fourth-generation launchers may appear. If engineers have their way, fourth-generation launchers will not resemble any of the rocketships in spaceflight history. They will not look like pointy-nosed V-2 rockets or advanced-technology spaceplanes. They could resemble one of the most familiar shapes from the popular culture of spaceflight. They may be round, disk-shaped objects, shaped like—would you believe—flying saucers.

With fourth-generation launchers, engineers hope to escape finally from the restrictions of chemical fuel. Some of the possibilities under investigation are quite exotic. The Beamed Energy Launch Vehicle concept uses laser or microwaves to propel space vehicles into orbit. Laser operators on the ground shoot a concentrated beam of light at the bottom of a spacecraft disk. Intense heat causes the air adjoining the disk to explode, propelling the spacecraft upward. Engineers at NASA's Marshall Space Flight Center are investigating "space elevators." The concept relies upon a 22,000-mile-long tether to lift humans from a high-altitude transfer station to an orbiting geostationary terminal.

Humans on the surface of Earth live at the bottom of a "gravity well" and an ocean of air. The speeds required to attain orbit—let alone travel to distant destinations—demand exceptional machines. For the last half of the twentieth century humans possessed rockets that could place payloads and people in orbit, but without the dependability and low cost with which objects routinely move across the Earth's face. Automobiles and aircraft have progressed from fragile, dangerous contraptions to relatively robust, comfortable machines in which lives are entrusted daily. The principal challenge for rocket engineers in the first half of the twenty-first century will be the development of similar machines—safe, reliable, low-cost vehicles operating between the surface of Earth and the edge of the space frontier.

Artist's rendering of a Microwave Lightcraft, about 2050.
Looking like an alien spaceship, the Microwave Lightcraft is an unconventional concept for delivering payloads to orbit. Microwaves beamed from a ground station or solar power satellite provide the energy necessary to power the spaceship, which employs exploding air molecules and pulse jet thrusters to create lift. Only a small amount of onboard propellant would be required for orbital maneuvers.

9015
U.S. AIR FORC
FE-158
RESCUE
DANGER
WARNING

05 : The Human Element

* * * * * *

preceding pages

Humans in space.

In 1959 NASA selected the Mercury 7 astronauts, the first Americans to fly in space (left). Scott Carpenter, Gordon Cooper, John Glenn, Gus Grissom, Wally Schirra, Alan Shepard, and Deke Slayton (left to right) were test pilots trained in high-performance jet aircraft. The ultimate twenty-first-century objective for advocates of human spaceflight is the exploration of Mars, projected for about 2030. In an artist's rendering (right), the first astronauts to explore Mars stand in front of an ascent vehicle waiting to return them to Earth and salute the people and nations of the world that made the journey possible.

Part of the "space gospel" motivating both American and Soviet/Russian space efforts from their creation has been the belief that humans will live and work in space. Space, in other words, is not just for machines. Through advances in technology, gospel promoters say, humans will leave their earthly home, travel to the Moon and planets, and live there.

In 1959 NASA officials introduced the Mercury 7 astronauts, the first Americans chosen to fly in space. They were test pilots, willing–in the words of writer Tom Wolfe–to go up in dangerous machines, put their lives on the line, and exhibit the "right stuff." Astronauts became synonymous with the U.S. space effort, and most Americans could not imagine space exploration without them.

Early experience with orbital flight led the Soviet Union to deploy the Mir space station in 1986, the first effort to establish a permanent human presence in space. In the post–Cold War era, the United States and Russia along with other nations of the world agreed to construct a more elaborate International Space Station and staff it permanently during the first decade of the twenty-first century.

What will humans have to learn in order to make space their home? It's a pretty hostile environment out there. The desire to establish a human presence, so much a part of the early vision, will provoke twenty-first-century pioneers to pursue unusual initiatives.

The Case for Humans Humans have long expressed curiosity about the solar system and the universe at large. Prior to the twentieth century, people had no opportunity to explore such things except through ground-based telescopes and works of fiction. These contrivances inspired scientists and engineers to develop technologies capable of launching humans and their machines into space. The first spaceflight pioneers translated centuries of dreams and observations into a practical exploration program.

Nearly all of the early pioneers believed that humans in spaceships would soon explore the solar system. Many struggled relentlessly to make that belief come true. They worked hard to convince a skeptical public of the desirability of this endeavor and to enlist tax-supported funding. To build public interest and justify government support, they turned to five central themes: national prestige, national defense, economic competitiveness, human destiny, and survival of the species.

At first Soviet and American politicians decided to invest in human spaceflight primarily because of national prestige. Space exploration became the primary means by which superpower leaders sought to demonstrate the superiority of their respective systems to uncommitted nations in the third world as well as to citizens at home. In that sense, peaceful space accomplishments like the landing on the Moon directly contributed to national defense. The image of Americans standing on the Moon was as important to the outcome of the Cold War as the military reconnaissance satellite. The Moon landings demonstrated that the United States possessed the capability to prevail in a struggle determined on a technological field.

During his 1960 presidential campaign, John F. Kennedy likened control of space to control of the seas in prior times. "If the Soviets control space," he warned, "they can control Earth, as in past centuries the nation that controlled the seas dominated the continents." To secure its position as a military superpower, the United States has maintained the most aggressive space program in the world.

Orbital flights and trips to the Moon were not undertaken primarily for scientific understanding, although scientific knowledge was attained. More important, the decision of U.S. leaders to be "first in space" provided a symbolic commitment to the assumption that the economy of the future would be based on science and technology rather than smokestack industries. In the beginning, spending on human spaceflight was used to revitalize the economy of the American South. Government installations in Florida, Alabama, Mississippi, Louisiana, and Texas created the economic seeds from which high-tech industries sprouted. The spaceflight program was also used to strengthen university programs in science and engineering and to interest bright, young students in those fields. Even though the Cold War ended, the commitment to spaceflight as a means of maintaining superpower status and a high-tech economy remains strong.

These are not arguments for a spaceflight program dominated by humans, however. National prowess and technological achievement can be demonstrated by space telescopes and planetary missions on which no humans ride along. When asked to explain if there was anything humans could do above the surface of Earth that instruments could not accomplish, one scientist replied, "Yes, there is. But why would anyone want to do it at such a high altitude?" As space exploration progressed, and the Cold War waned, scientists became increasingly skeptical of the value of human spaceflight.

below left

Maxime A. Faget, 1967. Wearing his characteristic bow tie, Max Faget inspects the Apollo command module. Trained as an engineer at Louisiana State University, he helped design every human spacecraft that NASA built during the twentieth century.

above

Astronaut Alan Shepard, 1961. The first American to enter space, with a 15-minute suborbital flight on May 5, 1961, prepares for launch in his Mercury "Freedom 7" space capsule.

"We set sail on this new sea because there is new knowledge to be gained and new rights to be won, and they must be won and used for the progress of all people. For space science, like nuclear science and all technology, has no conscience of its own. Whether it will become a force for good or ill depends on us. . . . Space can be explored and mastered without feeding the fires of war, without repeating the mistake that man has made in extending his writ around this globe of ours."

—President John F. Kennedy, 1962.

Faced with such developments, advocates of human flight shifted their ground. They reemphasized the importance of space exploration as an arena in which participants could renew the human spirit. "We choose to go to the Moon and do other things," said President John F. Kennedy in 1962, "not because they are easy, but because they are hard." Astronaut John Glenn, the first American to orbit Earth, pointed to space as "the modern frontier for national adventure." It is the medium, he said, "in which to test our mettle." Novelist James Michener, after writing a fictional account of the Apollo space program, warned that "a nation that loses its forward thrust is in danger." Editors titled his statement "Manifest Destiny," an unfortunate reference to the efforts of Europeans to colonize the American continent without regard to the needs of people who already lived there.

Footprints on the Moon, 1971. The Moon landings from 1969 to 1972 fulfilled a major objective in the U.S. human spaceflight program. A lunar rover allowed astronauts David Scott and Jim Irwin to range over the lunar surface more than 15 miles from their spacecraft. American astronauts left footprints and discarded equipment behind.

Manifest destiny conjures images of conquest and exploitation. As such, it is not a universally appealing rationale for human spaceflight, although it powerfully motivates the people who proclaim it. Advocates thus turn to the ultimate rationale for human spaceflight. It is necessary, they say, for the survival of the species. No long-lived technological civilization can remain forever on its home planet and survive, astronomer Carl Sagan proclaimed in his book *Pale Blue Dot.* The Sun will explode and die; an asteroid will batter civilization. The eventual choice, Sagan said, "is spaceflight or extinction." Humans became the dominant species on Earth by their willingness to migrate out of Africa and settle terrestrial regions where they could not live without the accoutrements of technology. Sagan fully expected that the basic human drive to settle new lands would continue in space.

Reconsidering the von Braun Paradigm All of these rationales found expression in the public relations campaign mounted by spaceflight pioneers during the 1950s. Wernher von Braun, Willy Ley, and Arthur C. Clarke proposed plans of remarkable consistency. No single person had more influence in defining this adventure than von Braun. Dominating the promotion of spaceflight in the 1950s, von Braun outlined an integrated space plan for human exploration that contained the following ingredients:

1. Earth orbital satellites to test technological requirements for human flight
2. Earth orbital flights by humans
3. reusable spacecraft for travel to and from Earth orbit
4. a permanently inhabited space station from which to observe Earth and launch expeditions to the Moon and beyond
5. human exploration of the Moon
6. human expeditions to Mars

The final step in the von Braun paradigm, the only one not accomplished or under way as the twentieth-first century began, is a human expedition to Mars. Many space advocates continue to press for this goal. In 1998, Robert Zubrin established the privately organized Mars Society to help build public support for such a mission. As one of the most persistent advocates for this incredible adventure, Zubrin insists that the exploration and colonization of Mars will inspire new technology, new science, vast creativity, and individual inventiveness among the people who undertake it. Humans with government funding may travel to Mars in the first half of the twenty-first century. If they do so, it will be for the more inspirational reasons that motivate human spaceflight. It will be done for reasons of national prestige and human destiny.

Such a mission will be exceptionally hard. Only about 40 percent of the American public surveyed in opinion polls consistently support using tax funds to send humans to Mars. Inevitable difficulties will undermine political support. If humans go, however, they will not go because it is easy to do. In the words of John F. Kennedy, they will go because it is hard.

Alternatively, humans may adopt goals for the next fifty years that are less ambitious, still inspiring, and certainly more practical from a political standpoint. Three of these are eminently achievable and will most assuredly take place in the near term. First, as an extension of normal airline operations, humans could travel between cities on Earth by traveling above the atmosphere, experiencing spaceflight for limited periods of time.

Second, the potential for extensive activity in low Earth orbit is excep

Wernher von Braun and Rocco Petrone, 1967.
As a U.S. Army and NASA employee during the mid-twentieth century, Wernher von Braun (right) helped define the principal steps for the human exploration of space. Here he discusses procedures with Rocco Petrone (left), a launch director at the Kennedy Space Center.

tionally bright. If this succeeds as planned, the International Space Station will establish a permanent human presence from which privately financed industrial plants could arise. Third, humans could return to the Moon, explore its most interesting features, and establish a permanent human presence. Permanent stations on the Moon would look very much like the internationally supported research stations located on the Antarctic continent near the South Pole.

In addition to these initiatives, humans may travel throughout the solar system in ways unimagined by the first pioneers: that is, by not physically going at all. Using the power of remote sensing, humans could establish a virtual presence on all the planets and their moons, and thereby experience these sites without leaving the comfort of their homes on Earth.

All such efforts would substantially modify the von Braun paradigm. Humans would not progress naturally toward the colonization of Mars, although extensive exploration with robotic machinery could take place, and human visits might someday occur.

Boarding the Spaceplane During the administration of President Ronald Reagan, senior government officials began to discuss the possibility of developing an "Orient Express," a hybrid air- and spaceplane that could carry ordinary people between New York City and Tokyo in about one hour. How is this possible? Actually, the concept is quite simple: develop an aerospace plane that can take off like a conventional jetliner from an ordinary runway. Flying supersonically, it reaches an altitude of 45,000 feet, where the pilots start scramjet engines, a jet technology that has the potential to push jetcraft to hypersonic speeds. The spaceplane rises to the edge of space and darts to the opposite side of the globe, where the process is reversed and the vehicle lands like a conventional airplane. It never reaches orbit, but it flies in space. The experience is similar to orbital flight, except for the shorter time.

The spaceplane concept has enormous promise and should become reality within the first half of the twenty-first century. Spaceplanes promise passengers

Artist's concept of a hypersonic spaceplane, 1986.
Part of the spacefaring dream is the hope that ordinary people will fly into space. During the 1980s, U.S. government officials began investigating the technologies necessary to build a hypersonic passenger carrier. Using turbojet and scramjet engines, the spaceplane would fly through the atmosphere to the edge of space, where it would cruise at hypersonic speeds between destinations on Earth.

Photographs of the Hubble Space Telescope, 1999.
The Hubble Space Telescope (bottom) was designed so astronauts could maintain it, vastly extending its life and capability. In 1999, shuttle astronauts C. Michael Foale (top left) and Steven L. Smith (top right) captured the telescope and replaced gyroscopes, electronic components, and part of the solar power system.

an opportunity to travel around the globe with greater speed and ease than any jetliners provide today. In the process, these passengers will become the first space tourists.

The cost of such flights will be high, without question. Technology within reach would allow corporations to build passenger spaceplanes and sell tickets for about $100,000 per seat. Does a market sufficiently robust to support this effort exist? Market studies suggest that at least a hundred thousand passengers a year would fly spaceplanes at the price noted here. That is a $10 billion-per-year business. It would grow in size and become less expensive as technology progresses.

The most attractive part of spaceplane travel at first will be its novelty. Like flying on the Concorde between Europe and New York City, it will not sustain itself solely as a practical means of transportation. Instead, bragging rights for having flown at hypersonic speeds will sustain much of the effort–that and the most exciting part of the flight, weightlessness. As the spaceplane travels at the edge of the atmosphere, passengers will experience about twenty minutes of free fall. Floating within the cabin, they will peer out of ports to the blackness of space and the blue-green Earth below. Given the technical definition of the term, they will qualify as astronauts–persons engaged in spaceflight.

Passenger service of this sort offers exceptional promise as an alternative means for financing space ventures. No longer dependent on government financing, space entrepreneurs may be able to raise funds for human spaceflight through the private sector. This is a critical step in opening the space frontier to ordinary people, thus helping to realize the promise that anyone can fly (with enough money).

Building an Orbital Research Park Exploration advocates want to build space stations as the primary destination for winged spaceplanes. For most advocates, building permanently occupied space stations was an early step in the spacefaring dream–not something that would occur after fifty years of flight. The actual space station has a tortured history. In 1984 President Ronald Reagan proposed the construction of a space station at a cost of $8 billion in 1984 dollars. NASA agreed to complete the facility in ten years, by 1994. Redesign followed redesign; costs soared. A multinational consortium, led by the United States, did not begin assembling the International Space Station until 1998. Dated from the year of its approval, the station will take more than twenty years and more than $50 billion to complete. Building and financing the station have proved far more difficult than its advocates imagined, retarding the progress of human spaceflight.

Once the station is permanently occupied, it could revitalize the space-

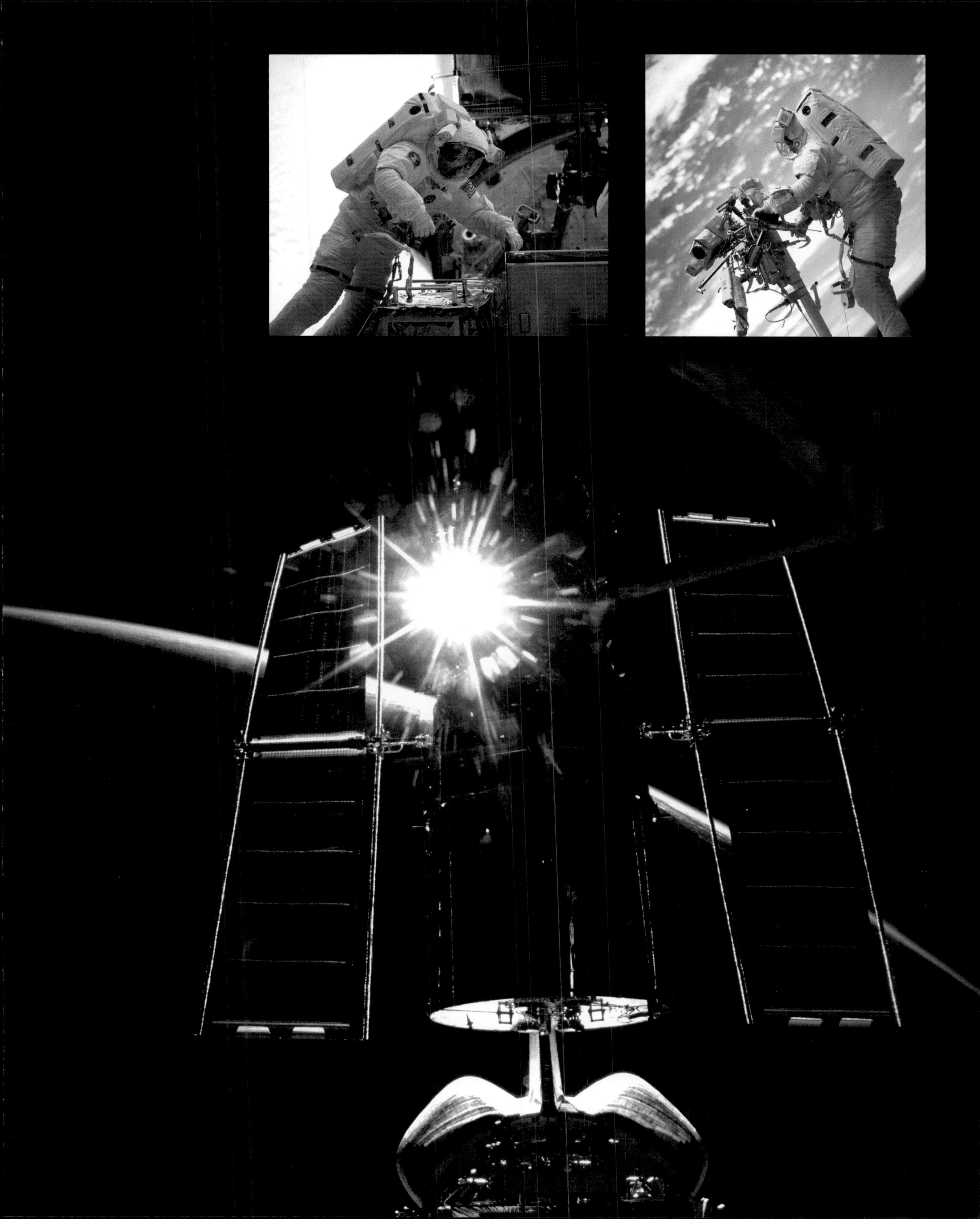

faring dream. "There will be an International Space Station," predicts John M. Logsdon, dean of space policy analysts. Barring a major catastrophe such as another shuttle accident, construction of the space station will be accomplished. This will be a remarkable achievement, given the technical, financial, and political obstacles involved. The U.S. House of Representatives came within a single vote of canceling the effort in 1993. Space organizations from different nations struggled to overcome cultural differences on a high-technology undertaking. Logsdon believes that the space station will provide the model for tax-financed human activities in space, creating a de facto world space organization with members prepared to cooperate on future missions. One hundred years hence, humans may well look back on the building of the station as the first truly international endeavor on the space frontier.

Assembling the International Space Station, 1998. Astronauts join the station's first two elements, the American "Unity" module and the Russian "Zarya" module, during a space shuttle Endeavour mission.

Once functioning in space, the station should energize the development of private orbital laboratories. Such laboratories would travel in paths near the International Space Station. The high-tech tenants of this orbital "research park" would take advantage of the unique features of space. Vessels traveling through space experience a measure of weightlessness calculated in micro g. One micro g is equal to the force of gravity felt on the surface of Earth divided by 1 million. Orbiting laboratories achieve microgravity levels between 1 and 10 micro g. This permits research not possible on Earth in such areas as materials science, fluid physics, combustion science, and biotechnology.

Consider protein crystal growth, the most important research tool for biotechnology in the twenty-first century. Scientists use x-ray diffraction of crystals to determine the structure of proteins. Such knowledge allows them to alter proteins or synthesize them in mass form. Protein crystals produced on Earth under gravity are malformed. Those produced in microgravity are nearly perfect. Scientists want to study protein crystals formed from a wide variety of substances, from insulin for diabetic treatments to plant enzymes targeted by fungicides.

Advocates hope that discoveries made on the International Space Station will encourage the development of privately financed modules in close proximity to the "anchor" station. These high-tech research parks would accelerate research on a wide variety of products. Most of the modules would be run by machines—human movement jars the delicate nature of microgravity research. The modules, however, would require periodic attention from technical teams. Recruits on these teams would reside in pressurized modules connected to the International Space Station or flying nearby. They would become the first routine space-farers on corporate payrolls, spending months away from home in a hostile environment.

If scientists and technicians can travel to habitat modules in the heart of this orbital research park, why not dedicate housing to those traveling for pleasure? Indeed, that was the premise of MIR Corporation, the international company that planned to use the Russian Mir space station as a hotel for high-priced vacations. With a research capability in orbit, can tourism be far behind? Probably not.

NASA has placed an order for at least one habitat module to be built for the International Space Station. One design calls for a school-bus-size aluminum cylinder; another calls for a three-story, inflatable structure shaped like a barrel. Inside, crew members will find phone-booth-size sleeping quarters, showers, a ward room, a cooking galley, and a health-care clinic. With the knowledge gained from producing habitat modules for the International Space Station, aerospace workers could produce additional models for space tourism. Tourist modules would require less instrumentation but more portholes for looking around.

If the promise of the International Space Station is achieved, space tourists may be taking trips to habitat modules–hotels in space–by 2020. Market studies suggest that more than one thousand people per year are willing to spend $1 million each for a weekend in space. This produces a billion-dollar-per-year business with significant potential for growth. As the cost of space vacations drops to $25,000 per person, the number of potential customers rises to seven hundred thousand annually, representing a revenue stream of $17.5 billion per year. That is larger than NASA's annual budget.

Assembling the International Space Station, 1998.
Tightening bolts becomes complex in microgravity as astronauts Robert Cabana, Nancy Currie, and cosmonaut Sergei Krikalev make "Zarya" operational (top). Johnson Space Center controllers (above) oversee orbital operations.

Operations centered on the International Space Station could open space to humans in much the same way that forts on the American frontier forged links between scientific curiosity and capitalism. As this occurs, the role of the government will become less dominant as private entrepreneurs fill the space near Earth. NASA will continue its research activities and deep-space exploration. Widespread human spaceflight near Earth, however, would become the province of the commercial sector in the first half of the twenty-first century.

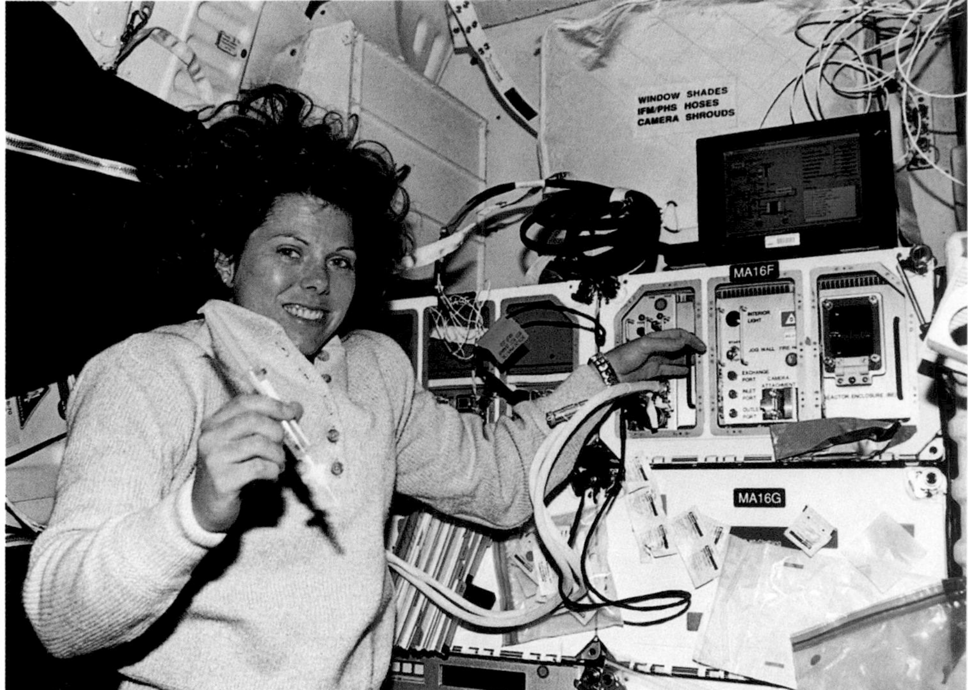

Microgravity research on orbit, 1995 and beyond. Using the International Space Station, shown in an artist's rendering (above), advocates of human spaceflight hope to develop products that will revolutionize life on Earth. In the precursor to the necessary experiments, shuttle astronaut Mary Ellen Weber (right) was photographed using a bioreactor to produce cells that grow only under weightless conditions.

Occupying the Moon During the next half century, a permanent human presence could be established on the Moon. It is no longer hard to get there. All of the technology needed to land and return is known. The space shuttle accelerates its passengers to 17,500 miles per hour. The passengers then transfer to an interplanetary spacecraft that takes them to the Moon. The International Space Station, or its successor, can be reconfigured to hold a lunar transit vehicle. Alternatively, a transit vehicle derived from station technology could be placed in a permanent orbit cycling between Earth and Moon. Such an endeavor would require a moderate investment, but the results may be astounding.

With this routine accessibility, humans could build an outpost on the Moon. The Moon has an abundance of materials from which to create a self-sufficient base. Near the Moon's South Pole, ice has been detected. From ice, humans could create water, oxygen, and hydrogen. The latter could be used to produce rocket fuel and generate electricity. Solar rays would provide an additional source of energy for the half month that the Sun faces that section of the Moon.

Why return to the Moon? This is a critical question, especially since humans have already been there. Why not press on to Mars? The advantages of lunar development are considerable.

First, the Moon lies only three days' travel time from Earth, whereas a three-year commitment is required for a round-trip voyage to Mars. A medical emergency involving a crew member on the Moon can be dealt with more effectively than one encountered during a journey to Mars. If the patient can be stabilized, he or she can be returned to Earth within a few days. In the event of any other crisis on the lunar base, help is only three days away. Maintaining a lunar base is similar to taking swimming lessons in the shallow end of the pool. One fully intends to venture into the deep end and jump off the high dive, but that comes after considerable practice and experience.

Second, the Moon offers what engineers call a test bed for trying the technologies required for more extensive space exploration. Life support systems and electric power plants, all necessary to sustain human life on extended voyages, can be tested on the Moon. The Moon provides an in situ demonstration project for the technologies necessary to survive the dangers of deep space. At the same time, the Moon offers a test bed for resource extraction. If humans are to explore Mars by "living off the land," lunar development provides a critical first step in the process.

Third, the Moon provides an excellent location for the advancement of astronomy, geology, and other sciences. An observatory must be established on the backside of the Moon. Such a platform for astronomy offers many of the advantages of Earth-based observatories, such as stability and human access, with none of the liabilities of flying observatories such as the Hubble Space Telescope. The observatory's images will be crisp since there is no atmosphere on the Moon. Located on the backside of the Moon, the observatory will be shielded from the reflected light of Earth, as well as radio noise, thereby offering great clarity for capturing images of astronomical objects.

Finally, a lunar base would expand the tradition of international cooperation established on the International Space Station. More nations would join the endeavor. People from many cultures would learn how to coordinate private firms and public agencies. Government officials will have to write agreements regulating access to the Moon and establishing property rights for entrepreneurs that make claims there. Who, for instance, will own any ice discovered at the lunar poles? Given the cost of transporting resources from Earth, lunar ice will be worth more than gold.

An international lunar base may help people on Earth live together in greater harmony. The Moon with a permanent human presence would become another Antarctica–an international protectorate with scientists from many nations seeking to understand space as they learn more about humanity.

Wiring the Solar System After the Moon, exploration teams will reach for the planets. Perhaps a few humans will travel to Mars in the next fifty years as part of a scientific expedition; perhaps not. Regardless of that, millions of people can travel through telepresence, and not just to Mars, but to the whole solar system. Scientists and entrepreneurs alike want to establish a virtual human presence throughout the solar system, using an assortment of low-cost spacecraft and communication transmitters. Within fifty years, humans could "wire the solar system." Landers and orbiting satellites could beam back to dot-com sites on Earth the latest images and weather reports from Mars and other locations.

In the summer of 1997, a NASA team from the Jet Propulsion Laboratory placed a small lander and its Sojourner rover on an ancient Martian floodplain. Millions of people used their computers to call up images returned from the Pathfinder lander and the semiautonomous rover via the World Wide Web. Twenty Pathfinder mirror sites recorded 565 million hits worldwide during the period from July 1 to August 4, 1997. The highest volume occurred on July 8, when a record 47 million hits were logged. That is more than twice the number received by the next most popular event, the 1996 Olympic games in Atlanta, Georgia.

NASA engineers and their industrial counterparts are working to develop miniature, low-cost spacecraft powered by inexpensive propulsion systems. The Pathfinder project was part of this "faster, better, cheaper" initiative, pioneered by NASA administrator Daniel Goldin in 1992. Advocates envision a future in which cameras no larger than microchips can be mounted on spacecraft the size of insects, powered by low-weight ion propulsion engines. Many

Artistic depictions of lunar activities, after 2020. Astronauts break ground for an observatory on the far side of the Moon (left). The lunar soil contains resources such as oxygen, iron, aluminum, and silicon that could be refined to help sustain a lunar base (below). Once the International Space Station is complete, advocates of human spaceflight will press to return to the Moon.

such spacecraft could be mounted on a single launch vehicle and shot to various destinations, where they would provide a virtual human presence from Mercury to Pluto. Their very low cost would make commercial funding possible.

Much of the data generated by such a system will be used for educational purposes. It is not difficult to envision that within the first half of the twenty-first century robotic explorers will be permanently stationed at the most interesting planets, moons, and asteroids of the solar system, sending back continuous data and images accessible to anyone with an Internet connection. Science and math classes around the globe would access this data for lessons and perhaps even perform analysis. The distribution of data for analysis by ordinary individuals has already been done on a limited basis by the SETI Institute. Participants receive data from radio telescopes searching for extraterrestrial civilizations and use home and office computers to scan the signals. Results are returned automatically to the SETI@home project. Anyone whose computer identifies a pattern that proves to be an actual signal will be named co-discoverer when the news is announced.

Eventually humans will learn how to create virtual expeditions. Amateur explorers wearing specially designed helmets will experience the thrill

"To answer the question 'Humans or robots?' one must first define the task. If space exploration is about going to new worlds and understanding the universe in ever increasing detail, then both robots and humans will be needed. The strengths of each partner make up for the other's weaknesses. To use only one technique is to deprive outselves of the best of both worlds: the intelligence and flexibility of human participation and the beneficial use of robotic assistance."

—Paul D. Spudis, lunar scientist, 1999

of exploration by "walking around" real sites on the Moon and Mars. But they will do so on Earth, using images transmitted from in situ spacecraft, without exposing themselves to the dangers of radiation and vacuums. In its own way, this would fulfill that part of the space gospel that predicts a day when anyone can travel into the void.

How will people fifty years hence view the first human expeditions into space, including the first landings on the Moon in 1969? Advocates of human exploration view the astronauts who circled Earth and landed on the Moon as people akin to fifteenth-century seafarers like Christopher Columbus, the vanguards of sustained human exploration and migration. Alternatively, those first space ventures may prove more like Leif Erickson's voyages from Scandinavia several hundred years earlier, an exploratory dead end.

The answer to this question depends upon the ability of humans to adapt their visions to the reality of space. Early visions of human exploration were entertaining, but they were rooted in a relative lack of understanding about the nature of the Moon and planets and a view of colonization outmoded even at that time. The tradition of human spaceflight may continue, but in ways not envisioned by the early pioneers.

Robots in space. Throughout the history of space exploration, automated satellites have tested the environments into which humans later followed. Scientists on Earth designed planetary rovers (far left) and sent one to Mars in 1997 (center). Sometime in the future, humans may venture to Mars and retrieve the frozen, windblown Sojourner rover (below).

06 : The Commercial Space Frontier

"We have outgrown the 'gee-whiz' factor that once fueled the space program. The litany of firsts, farthests, and fastests—so inspiring in real time—is history. Today, the Earth-orbit frontier has been mapped and charted. The benefits of satellites are well defined. The basic technologies are mature. The accessibility of space and demand for its products make it ripe for entrepreneurs."

—Timothy W. Hannemann, General Manager, TRW Space & Electronics Group, 1995.

Want to make money in space? The commercial space sector may well become a trillion-dollar-per-year activity by 2050, as routine and profitable as shipping and telecommunications are today. If this happens, watch it begin with telecommunications, move through mining and manufacturing, explode into space tourism both real and virtual, and expand through an aviation industry that learns how to create "seamless" flight—vehicles that fly in both air and space.

In the wake of this new commercial frontier, NASA and other government agencies will lose their place as the dominant operators in space. Exiting the truck-to-space business sometime after 2005, NASA sells the space shuttle fleet to private operators. The International Space Station is turned over to private industry after 2010. Future space activities are governed by the fundamental belief that the business community and private entrepreneurs can conduct routine operations better than government agencies. NASA performs cutting-edge research and launches scientific expeditions; private corporations do everything else. "Everything else" becomes the largest share of space activity. In this environment, the possibilities for making money in space through private activity expand substantially.

preceding pages

The marketing of space. Following the first landings on the Moon, portrayed in a 1961 Chesley Bonestell painting (left), space advocates anticipated that mining opportunities would spur lunar development in ways similar to gold rushes on the American frontier. In a 1980 painting (right), astronauts are depicted overseeing a lunar mining site and removing helium 3 for transport back to Earth. As of 2000, mining operations on the Moon were still decades away.

opposite

Research park in space, projected for 2005. Entrepreneurs from Spacehab, Inc., plan to attach a module to the International Space Station, where they will conduct microgravity research on a fee-for-service basis.

A Risky Business Optimists believe that private entrepreneurs will make money in a variety of space industries. These include communications, navigation, remote sensing, package delivery, space entertainment, space tourism, resource extraction, launch services, precision agriculture, and microgravity manufacturing. The process of converting business concepts into practical profits, however, is a challenging one. What looks good on paper often fails in practice. Consider the case of telecommunications.

In 1945, Arthur C. Clarke proposed the use of orbiting satellites as a means of transmitting television and microwave signals around the world. In 1962, NASA launched the privately built Telstar satellite, the first orbital device to relay television and telephone waves. In the years that followed, space-based telecommunications grew into a multibillion-dollar industry. In 1987, twenty-five years after the launch of Telstar 1, three Motorola company engineers proposed a space-based system that allowed telephone users to communicate anywhere in the world. It was a wonderful concept: An oil company executive, for example, standing on an isolated platform in the North Atlantic ocean places a phone call to Texas with a handheld telephone. The telephone signal rises to a constellation of satellites in low Earth orbit, which switch the signal to the person being called.

Satellite launches began in 1997. Iridium, a separate company incorporated by Motorola, went public with an initial offering of $350 million. In

1998 the company completed its constellation of sixty-six satellites. It began service on November 1, 1998, and filed for bankruptcy the following year.

Iridium was a debacle of enormous magnitude. The system was too expensive to compete in the marketplace. The benefits of making a wireless telephone call from a region not served by ground-based antennas did not outweigh the price Iridium charged. Iridium executives encountered technical difficulties. Users in large cities could not maintain access to the constellation of fast-moving satellites when tall buildings blocked the way. Corporate planners misjudged market growth for voice communications. Higher profits flowed to providers of Internet data and media information, so much so that those providers offered voice communications as a "free" commodity. Data transfer technology changed in unexpected ways. Makers of fiber-optic cable systems learned how to market their ground-based product at prices far below those charged by space-based satellites.

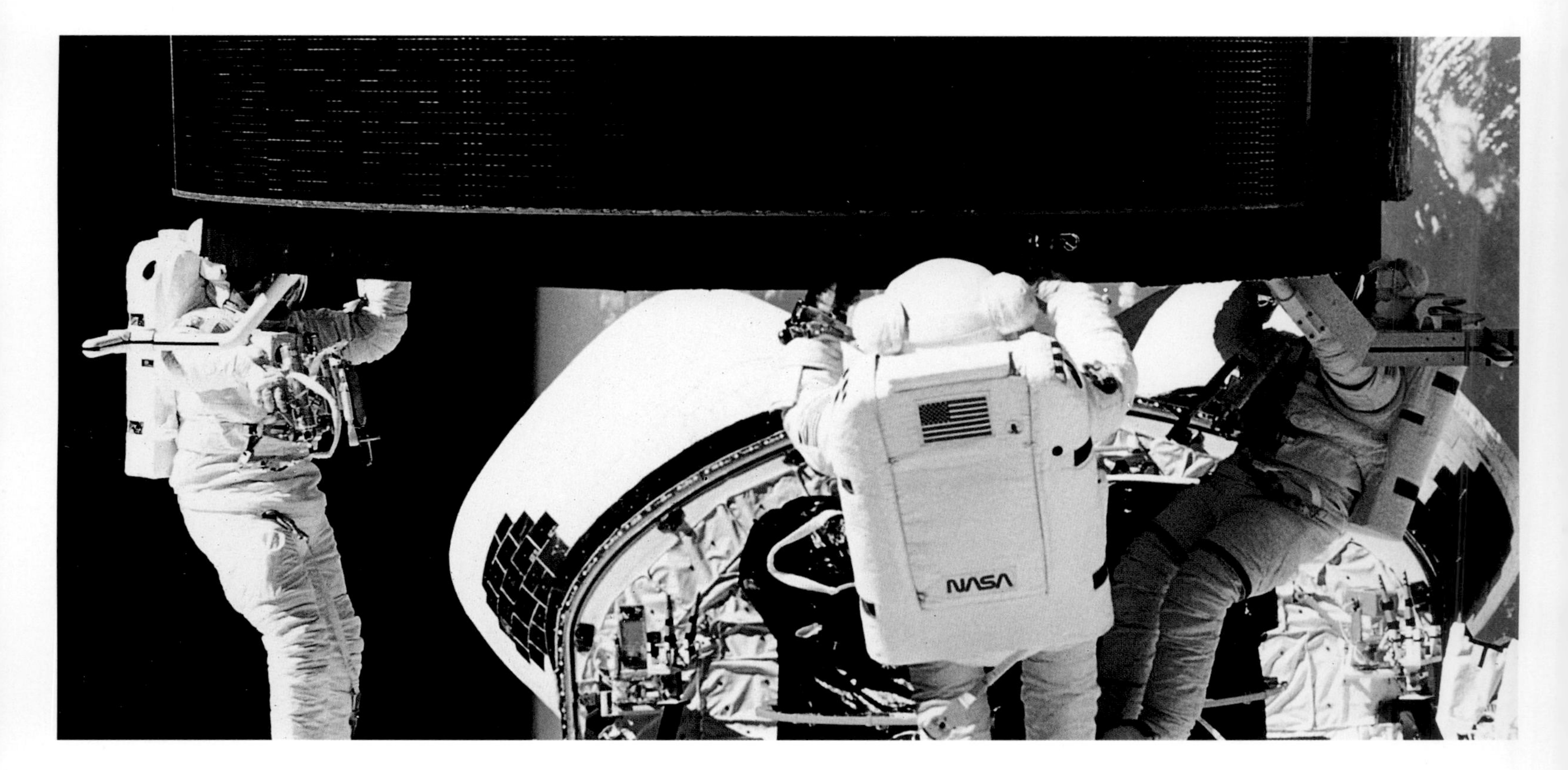

Photograph of astronauts capturing a communication satellite, 1992.
Satellite telecommunications has grown into a multibillion-dollar-per-year industry. Three crew members from the shuttle Endeavour retrieve and redeploy an Intelsat-VI communications satellite, which had been stranded in an unusable orbit since its launch two years earlier.

Space commerce is a high-risk undertaking. Four out of five companies entering the space commerce market fail and produce no profits at all. Most successful companies do not produce profits until their fifth year of operation. Because space commerce is a risky business, fraught with technological change, many venture capitalists prefer to invest in sectors such as biotechnology and the Internet.

Government regulations compound this risk. Before 1990, governments conducted nearly all space activities. Policy making took place in "non-transparent" organizations, well removed from commercial scrutiny. Those organizations did not promote space commerce, and what space commerce they allowed was characterized by excessive regulation. The Department of Defense, for example, placed satellites on its munitions control list. An entrepreneur who wanted to market American-built satellites to a foreign country had to wade through a torturous permit process.

Given the heavy hand of government, most space activities with commercial potential were technology driven rather than market driven. Government officials undertook technologically feasible activities like military spy satellites and paid corporations to build the hardware. The only major space activity driven by markets was telecommunications.

In the early 1980s, aerospace executives came to Washington, D.C., and asked government leaders essentially to get out of the way. The corporate executives wanted to develop new space industries and take over tax-funded services such as weather satellites. To do this, they needed assured access to space and a space launch pricing policy that would allow them to prepare financial plans. Government leaders told the industrialists that the reusable space shuttle would be the primary means for delivering commercial payloads and that corporations could purchase as many flights as they needed. In 1986, the space shuttle Challenger exploded and grounded the executives' dreams. Subsequent commercial activities were difficult to develop and especially hard to finance.

Some industrialists are very optimistic about the future of space commerce. They envision a future in which communication, navigation, and imaging satellites will capture substantial market shares. They believe that rocket engineers will discover vastly improved launch technologies. Entrepreneurs would like to develop space tourism and space "hotels." Roughnecks would like to produce power, fuel, and consumables such as water for tax-financed bases on the Moon and Mars. Information Age geeks would like to extend the Internet into space, producing a "virtual" human presence throughout the solar system. Employing the unusual properties of a gravity-free environment, manufacturers would like to produce a host of new products in space–insulin crystals, computer components, strange metal alloys, and microcapsules the size of white blood cells that can kill cancer cells with bulletlike accuracy.

Translating concepts like these into profitable enterprises will be difficult. Pessimists believe it will not occur. In their vision of the future, government bureaucrats stifle space commerce through excessive regulation. International disagreements destroy the market for space-based information industries. Terrestrial communication devices pollute the spectrum needed for global positioning systems. Transportation accidents cause a collapse of the private insurance market for commercial launchers. Clever scientists create space-age products that can be manufactured on Earth. What few space industries remain are nationalized in all but name, and large government contracts determine the structure of space industries, as was the case at the beginning of the space age.

No one knows exactly what will happen. New commercial opportunities brim with risk. Payoffs may be huge, but so are the opportunities to fail. People uncomfortable with risk should probably stay away. This is not an activity for the faint of heart. If the commercial space frontier expands as optimists believe, this is how it will likely occur over the next fifty years.

Satellite Telecommunications Business firms engaged in satellite telecommunications, the first major space technology to find commercial application, have been generating billions of dollars of sales annually for more than a quarter century. Humans would be hard-pressed to live without communications satellites. Imagine a world without instantaneous news from around the globe. What would the Internet be like without global satellite connections? Perhaps some people would prefer these limitations, but probably most would not.

Several companies currently provide fixed satellite telecommunications services within the United States: GE Americom, AT&T, COMSAT, GTE, Loral Cyberstar, Teledesic, and Hughes Space and Communications.

Artist's concept of a single-stage-to-orbit launch vehicle, about 2025.
Space entrepreneurs believe that low-cost, single-stage-to-orbit launchers will make space competitive for a variety of commercial operations, including the telecommunications transmitter being deployed in this depiction. Such launchers would reach orbit without discarding lower stages.

Worldwide, space-based telecommunications is a $35-billion-per-year industry. Revenues from communications satellite users support an elaborate network of satellite manufacturers, launch services, ground stations, and providers of technical support.

Satellite telecommunications ventures have been very profitable, even though firms operating in this area have been forced to pay excessively high launch costs. Industry advocates believe that the revenues generated from satellite telecommunications will multiply tenfold by 2010. The key, they believe, lies in personal wireless communications. Geoffrey Hughes, a prescient observer of the business, called personal wireless telecommunications "the undiscovered country." The technology that powers cellular phones will be extended to computers, electronic data, images, navigational information, and Internet material. Much of this activity will be transmitted through satellite systems.

Skeptics fear that the market for satellite transmissions is threatened by fiber optics, especially in North America and Europe where information volume is high. Satellite transmission, skeptics believe, may be confined to niche markets that are remote or thinly populated. The battleground for market shares, skeptics insist, will be confined to emerging markets in areas like Africa and Asia. Experts on both sides of this issue believe that people worldwide want a plethora of personal wireless options. Wireless telecommunications could dramatically change the lives of people in emerging economies with historically poor telecommunications capacity. Experts disagree on whether the systems they need should run through satellites or fiber-optic cables.

Hughes insists that telecommunications satellites can do the best job. Why would any corporate executive seek to wire an emerging nation with fiber-optic cables when the same job can be done faster, better, and cheaper using satellites–and result in fully mobile telecommunications as well? As Hughes has said more than once, "We are all going to be connected but unplugged. This will happen. Depend on it."

Looking back to the period before the first telecommunications satellites were launched, it is hard to imagine how humans carried out work without them. From the midpoint of the twenty-first century, the same may be said of wireless communications.

"Who can predict what we will find as we proceed over the next years to investigate our Solar System and the stars beyond? Who could have predicted in 1990 all that we have learned since then about water on Mars, potential early life on Mars, oceans beneath the ice of Europa, planets around other stars, and the robustness and early origin of life on Earth?

So in the coming years, as my 17-year-old son would put it, other stuff might happen. When it does, let's be ready."

—Wesley T. Huntress, Jr., NASA associate administrator for space science, 1998.

The Transformation in Remote Sensing Entrepreneurs are also investing in satellites that can take images of Earth. Remote sensing, as it is called, is the "second great commercial space hope." An increasing number of companies are competing to provide commercial images to customers around the globe, selling pictures from space as sharp as those generated by military spy satellites. Entrepreneurs in this industry promise wondrous growth; this may or may not occur. In 1999, the Space Imaging Company launched Ikonos, the first high-resolution satellite owned by a commercial firm. Prior to that time, private companies sold images generated by government satellites. Entrepreneurs believe that privately owned imaging satellites will create an industrial sector as rich as space telecommunications.

This industry draws on past government support for military reconnaissance, meteorological, and Earth-resource satellites. Military spy satellites provide the basic technology by which images are taken and transmitted back to Earth. The U.S. Air Force launched the first maneuverable reconnaissance satellite in 1960 as part of the supersecret Keyhole line. Keyhole satellites took photographs of ever-increasing quality in visible light, infrared, and multispectral wavelengths. Multispectral scanning is a unique process that allows analysts to determine the substance of the material being photographed—for example, whether a structure is made of plywood or steel. NASA launched the first successful weather satellite, Tiros I, in 1960. The satellite provided astounding images of weather fronts, storms, and cloud formations. More important, it led to a long series of weather satellites that quickly became standard weather-forecasting tools throughout the world.

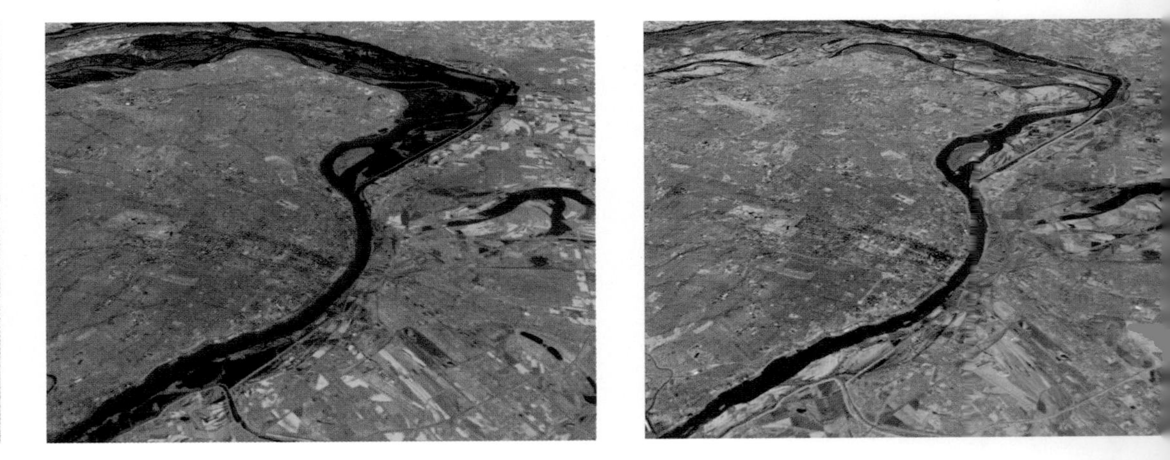

Images from Landsat satellite. Satellite images of the Mississippi River in and around St. Louis, Missouri, show the dramatic contrast between the normal flow of the river in 1991 (left) and the devastation of the 1993 flood (center). Artificial structures of concrete and asphalt appear dark gray or black. Reddish areas in an image taken a few months after the flood (right) indicate where water had receded, leaving barren land.

NASA launched the first Landsat satellite in 1972. It provided data on vegetation, insect infestations, crop growth, and land-use patterns. More Earth-resource satellites followed. Scientists used images to forecast worldwide crop yields. Other people used them to manage oil spills, aid navigation, monitor pollution, assist in water management, site new power plants and pipelines, and aid in agricultural development.

The benefits attributed to remote sensing satellites have been broad. Information about crop yields has significant economic value. Wheat yields are traded on special markets where producers and their customers guess the future direction of prices. Prices are volatile, moving up or down with variations in the size of worldwide yields. Information from remote sensing satellites helps to dampen speculative swings on an otherwise uncertain futures market.

Commercial remote sensing is a fledgling market with immense potential. It generates revenues of about $800 million per year worldwide. With the right tools, it could spawn new opportunities and industries worth as much as $8 billion a year within a decade or two.

Space-Age Farming Precision agriculture relies on methods that carefully tailor soil and crop management to conditions found in each field. It applies three related technologies to farming: satellite remote sensing, the Global Positioning System (GPS), and geographic information systems. Precision agriculture literally takes farming into the space age, since the first two technologies are based in space.

Each year farmers receive more information from satellites. They can download the information directly, or they can rely on commercial firms to do this for a fee. Remote sensing images distinguish crop species and locate stress conditions. These images, properly analyzed, can be mated to the Global Positioning System, allowing farmers to place precisely the right amount of fertilizer or pesticide at precisely the right location. This system enables more efficient, and ultimately less costly, agricultural practices that produce higher yields. Equally important, use of fewer chemicals through precision agriculture helps to preserve the environment.

Photograph of precision farming through space technology, 1998.
The device worn by this farmer holds sensors linked to the satellite-based Global Positioning System. Analysis of collected satellite images will indicate areas where soil and vegetation conditions need to be altered.

By knowing precise locations on a field, farmers can better analyze soil samples, adapt tillage practices, adjust seed distribution, and monitor crop yields. Through satellite observation, they can identify weeds, locate plant stress, check crop growth, and measure moisture effects. Farmers can compare results on different fields and changes from year to year. Satellite data can reveal discrepancies between predicted and actual yields.

Space-based technologies are just beginning to have an impact on farming. This is a brand-new industry. It requires firms to provide images in a timely fashion and entrepreneurs who can process the information into useful forms. Most of the technological components needed for precision farming are available now. Space-age industries that promote "farming by the inch" will expand in the twenty-first century. Farmers of the future may spend their evenings analyzing data and images from space using computer programs that calculate agricultural practices with space-age precision.

One-Hour Delivery Anywhere in the World Spaceplanes should appear in the twenty-first century. In addition to carrying passengers, they can deliver cargo, a potentially lucrative business.

Who would pay large amounts of money to deliver packages across the world in an hour or two? Most of the freight that travels through Federal Express and United Parcel Service would not qualify. Few will ever send Christmas packages this way, even late shoppers. A potential market for very rapid delivery does exist among high-value, low-weight parcels: electronic devices, organs for transplant, biomedical implements, and commercial documents.

When the space shuttle began operations, Fred Smith, the chief executive officer for Federal Express, offered to buy an orbiter to provide this very service, provided NASA would handle launch services on a reimbursable basis. Smith understood the potential market for fast package delivery that such a vehicle offered. "Based upon our research and investigation," he wrote in 1986, "a hypersonic aircraft would be economically viable." His customers, he said, were willing to pay for transoceanic speed.

Artistic renderings of spaceplanes, after 2020.
Planes capable of flying in both air and space could deliver people and packages to destinations around the globe—or above it. The spacebus (below) consists of a fifty-passenger orbiter riding piggyback on a high-speed booster.

NASA's space shuttle is too technologically fragile to perform this task, but the concept remains. Several experts estimate that annual revenues from fast-package delivery could exceed $1 billion per year. Once the service is under way, the market would increase by about 7 percent each year. Federal Express, United Parcel Service, and other companies have prepared plans for capturing this market. As just-in-time delivery becomes the norm, demand could explode. The new international economy requires seamless, quick (not initially inexpensive) delivery of high-priority cargo across the globe; spaceplanes hold the best promise for filling this niche in the transportation infrastructure in the twenty-first century.

Space Voyages for Fun and Profit Would you like to fly a satellite in a low orbit over the Moon, zoom in on features that interest you, and download images of the Apollo 11 landing site directly to your home computer? You may well have the opportunity before long. A commercial satellite to the Moon can be built for $30 million, with a comparable amount for launch and operations. Once in orbit around the Moon, such satellites could be used as a source of entertainment and education for a nominal fee. From theme parks or other selected locations throughout the world, the public could "fly" the spacecraft over the Moon, changing its orbital altitude and inclination to view interesting features up close. Those at home might not "fly" the orbiter, but they would be able to download photographs and video.

Carry this broadly attractive activity one step further. A corporation lands a rover on the Moon and allows the public to drive it across the lunar surface, investigating rock formations and Apollo landing sites. What would you pay for the privilege of driving a robot explorer on the Moon?

Several corporate leaders believe that consumers would pay a good deal. Among them are executives from LunaCorp, Applied Space Resources, SpaceDev, TransOrbital, and Astrobotics. They want to do it soon. The chief executive officer for Astrobotics wants to send satellites and rovers to the Moon within ten years, promising a 32 percent return to investors willing to provide $325 million in setup costs. The initial target would be Tranquillity Base, where the first astronauts to visit the Moon set down in 1969. Two rovers would land near the Apollo 11 site and explore for two years. The concept couples telepresence with the technology of immersive rides, creating the twenty-first-century equivalent of an old amusement park. The business also caters to amateur explorers, novelty hunters, and immortality seekers. Through the science of teleoperation, ordinary humans on Earth would fly the craft, drive the rovers, plan the routes, and guide the exploration effort.

Artist's concept of a lunar probe, about 2010.

Some corporate leaders believe that the general public will pay to experience space through high-quality images generated by satellites and rovers. The public could drive these vehicles from Earth in the twenty-first-century equivalent of the amusement park ride.

The commercial market for such activities is the entertainment dollar. Americans spend $500 billion per year on entertainment, and space entrepreneurs want to capture a portion of this market. For people with lots of money to spend, there may be space hotels. For a number of years, space enthusiasts have proposed hotels traveling in low Earth orbit or resting on the Moon. One proposal envisions a 101,000-ton structure moving 775 miles above Earth. After paying $94,000, guests board a private spaceplane bound for an orbital dock 160 miles high. From the dock, guests reach the hotel in a special elevator that ascends a 615-mile tether. British architect Peter Inston, who has designed hotels for Hilton International, proposes a lunar hotel larger than the MGM Grand, Las Vegas, a five-thousand-room domed structure powered by solar energy and utilizing water taken from lunar ice. In Inston's plan, special buses transport guests on low-gravity excursions across the lunar surface. One Japanese firm spent $3 million developing plans for lunar condominiums.

Such ideas, to say the least, are premature. The first orbital hotels will consist of space station modules housing six or seven guests and one staff employee. Still, entrepreneurs remain optimistic. "As the business grows, there

Artist's rendering of a space hotel, after 2020.
Entrepreneurs want to sell vacations in space. This elaborate structure, painted by Pat Rawlings, combines an entertainment center with a privately-run research facility. A solar tower provides electricity for tourists and scientists on board.

will be a demand for larger and more complex structures to be assembled in orbit and there will be almost no limit to the range of shapes and sizes that may be built," says Patrick Collins, vice-president of Spacetopia, Inc. Theme parks and religious temples may follow.

Mining the Asteroids Space entrepreneurs want to undertake one other important activity during the next fifty years. Like gold rush pioneers on the American frontier, they want to mine the solar system. Space enthusiasts have been talking about this type of activity since the space age began. In the past, they envisioned humans, assisted by complex machinery, mining precious commodities such as helium 3, which exists abundantly on the Moon but not on Earth and can be used as fuel for fusion reactors. They offered to extract raw materials for the construction of space stations and extraterrestrial bases. As on the American frontier, space advocates view resource extraction as one of the most readily attainable commercial activities for a space-faring civilization.

The first mineral sources in space likely will be near-Earth asteroids. While most asteroids are located in a belt between Mars and Jupiter, a few pass close enough to Earth to make mining feasible. NASA sponsors a constant watch for celestial bodies that might collide with Earth; participating scientists have discovered several near-Earth asteroids that contain useful metals. Scientists use telescopic spectroscopy, which analyzes light reflected from such objects, to determine their makeup. Two asteroids, known as 1985 EB and 1986 DA, contain nickel and iron. Both asteroids approach Earth at a distance about as close as the Moon. On a future approach, resource extraction could begin. Another, subsequent near-Earth asteroid, 1993 BX3, weighs several million tons and contains metal as well. Scientists estimate that two to three thousand such asteroids orbit nearby. Not all contain metal, however. Some are suitable only as sources of stone.

Mining a metallic asteroid is easier than mining the Moon. On an asteroid, pure metal can be extracted directly; it requires little processing or purification. In addition, asteroids contain small amounts of hydrogen, carbon, and water, all useful substances for space pioneers. Given the subzero conditions existing in space, hydrogen and water can be transported as frozen lumps with a simple shield to keep them shadowed from the Sun.

To mine asteroids, some engineers favor orbital alteration. Small rocket engines would deflect the object from its solar orbit and send it into an elliptic orbit around Earth. Others favor machines that attach themselves to the asteroid as it flies by Earth, extracting materials on the outbound voyage and dispatching the processed materials when the asteroid returns. To expedite this process, machines may mine frozen gases such as hydrogen to use as fuel.

In most cases, machines will conduct mining operations. Machine

Orbital Logistics, Inc
4
SPARK

NASA
D STATES
NASA
NASA

mining under conditions of very low gravity will be hard. Consider an asteroid consisting of a crumbly material such as silicate dirt embedded with nickel-iron granules. For such bodies, most experts favor strip-mining, in which machines scrape the asteroid's surface in search of desirable materials. But how does one hold the cutting edge of the machine against the ore? Harpoons or anchors could be driven into the surface. Alternatively, cables or a net might encircle the asteroid. Strip-mining would release tons of waste rock into orbit around the asteroid, creating a debris cloud that might cover the machine and interfere with mining operations.

Some engineers favor tunneling, a method used for centuries on Earth. Tunneling would prevent consumption of the entire asteroid, as the mining machine would follow only desirable veins of ore. The cutting machine might hold itself steady by pressing robotic arms against tunnel walls.

In spite of countless feasibility studies, the mineral exploitation of outer space is decades away. Initially, profits from space mining will not compete with the business of mining materials on Earth. In time, however, as space activities expand, this will change.

Commercial activities such as mining and entertainment create a special problem–who will own the property rights to the resources produced? For space activities funded through government agencies with tax-collected dollars, ownership is not an issue. The products produced are public goods, owned by the society at large. For corporations that must repay their investors, however, property rights are a prime concern. So long as the government treats space as a public good, no economic incentive to undertake commercial activities in space will exist. Until this issue is resolved, mining operations are not only unlikely; they are probably illegal.

Perhaps by 2050, members of the United States Congress will propose a lunar grant the size of Alaska, encompassing a polar crater of permanently frozen water and an adjacent mountaintop with a sunlit crest. Such a land grant would be worth a fortune today, even without an inexpensive way to get there. How much more would such a grant be worth once a privately owned company established a base on the Moon, with spaceliners going back and forth carrying paying passengers and commercial activities such as resource extraction?

As a legacy of the twentieth century, government practices created substantial obstacles to the commercial development of space. Sometime in the twenty-first century, that situation will change. Some government will pass legislation stating its intent to recognize and defend the validity of property claims by space-faring companies. Such laws will encourage entrepreneurs to pursue space-based endeavors.

Only then will the space frontier truly be open.

Artist's concept of a transportation depot around Mars, 2050.

Commercial activities may extend throughout the solar system. Businesses will provide basic materials like fuel and water and operate transfer stations like this one depicted by Robert McCall, often under contract with government exploration agencies.

4 ОКТЯБРЯ 1957 ГОДА
В СОВЕТСКОМ СОЮЗЕ
ОСУЩЕСТВЛЕН УСПЕШНЫЙ ЗАПУСК
ПЕРВОГО В МИРЕ
ИСКУССТВЕННОГО СПУТНИКА ЗЕМЛИ
ЭНЕРГЕТИЧЕСКАЯ СИСТЕМА СССР
МОСКВА
СТАЛИНГРАД
АСТРАХАНЬ
САРАТОВ

07 : Space Warfare

* * * * * *

preceding pages

Making war from space.

Sputnik 1 (left), launched in 1957, established an important principle of spaceflight: satellites flying over foreign territories do not violate the airspace of those nations. Both the United States and the Soviet Union deployed an extensive network of spy satellites to monitor each other's activities. This U.S. Defense Support Program satellite, shown in a 1994 artist's rendering (right), can detect missile launches and nuclear explosions.

opposite

Photograph of military installation, Groom Lake, Nevada, 1998.

Reconnaissance satellites take high-resolution pictures of top-secret sites. A Russian SPIN-2 spy satellite produced this image of the closely guarded U.S. military installation at Area 51, long thought by conspiracy buffs to house alien spacecraft.

Whenever civilizations move into new territories, military forces accompany them. Military forces often lead the way. Officers organize expeditions of discovery and, once settlers arrive, provide security and protect resources. In American history, the military played a significant role in securing the frontier. Meriwether Lewis and William Clark led a military mission designed to scout the landscape and contact the indigenous peoples who already lived there. The army on the frontier protected settlers, fought engagements, and escorted convoys. It also advanced the frontier by constructing outposts, roads, harbors, and waterways.

The military is involved on the space frontier in ways both traditional and strange. One popular concept holds that the military, as in the film *Independence Day,* will defend society from alien invasion. This is not likely to happen in the next fifty years. Others believe that military involvement will be minimal, since the United States and other nations are committed to the peaceful uses of space. This is naive. The military is already in space, and its presence will grow more pronounced as the century progresses.

During the twentieth century, military forces used space for eyes, ears, and crosshairs. The superpowers agreed not to place weapons in space, but they launched every other part of the gun. By deploying spy satellites, military forces gained the capability to select targets and guide "smart bombs" toward them.

During the twenty-first century, the use of space to protect the ground will grow more sophisticated. Military forces will learn how to defend their own satellites and disable hostile ones. They will be called upon to protect national assets in space, renewing pressures to create a military astronaut corps. Military forces may construct space facilities in a manner similar to the work of the U.S. Army Corps of Engineers. They may be called upon to defend the world from the most likely form of alien invasion—a comet or asteroid on a collision course with Earth.

Supporting National Defense Since the beginning of the space age, the United States military has used space to support defense activities on the ground. In the 1950s, President Dwight Eisenhower authorized the deployment of spy satellites that could pierce the iron curtain of secrecy shrouding the Soviet Union. Eisenhower feared that Soviet leaders would prepare for war—even launch nuclear missiles—without the United States detecting these activities. Determined to prevent a Soviet surprise attack similar to the one the Japanese launched against Pearl Harbor in 1941, Eisenhower sent high-altitude balloons and U-2 spy planes on missions across Soviet territory. Balloons and planes proved ineffective, and in 1960 the Soviet military shot down one of the spy planes and put the pilot, Francis Gary Powers, on trial.

Later that year, leaders of Project Corona launched the first U.S. reconnaissance satellite. A new era of intelligence gathering began. This highly classified reconnaissance effort acquired 3,000 feet of film with coverage of over 1.65 million square miles of the Soviet Union. It revolutionized the way in which the United States government collected foreign intelligence. Through spy satellites, the intelligence community received timely images that offered a synoptic view of much of the Earth's surface.

Corona was succeeded by a series of ever more sophisticated reconnaissance satellites and a continuous stream of data. Analysts reviewing satellite data counted enemy missiles, tracked troop movements, inspected military facilities, and verified weapon reduction agreements. Overseen in the United States by the National Reconnaissance Office, this top-secret effort informed the decision making of American governmental leaders throughout the Cold War. President Lyndon Johnson, in an unguarded moment, confessed that the cost of the spy satellite program paid for itself tenfold. Spy satellites allowed U.S. leaders to calculate the exact size of the Soviet military, he said, absolving them of the need to build defensive systems for protection against weapons that did not exist.

The first spy satellites took conventional photographs. They could not see through clouds, so scientists developed radar devices that did. Nor could they see tanks or other weapons hidden under camouflage, so military leaders commissioned infrared satellites that detected heat imprints. The first spy satellites could not distinguish between sophisticated jet aircraft and cheap ones that would not perform, so scientists developed multispectral scanning satellites that could. The satellites took multiple pictures in different wavelengths which, when analyzed, revealed whether an object was made of titanium, aluminum, or plywood painted to look real.

To warn against a Soviet ballistic missile attack, and thereby prevent one from happening, officials in the Department of Defense developed satellites with infrared photographic equipment that could detect missile launches. The satellites recognize heat signatures from rocket blasts, which then appear on special monitoring screens at military posts. Within seconds of a launch, analysts can identify time and place, missile trajectory, and final target.

Infrared photographic technology was a brilliant concept but it took years to develop. Leaders of the first effort, Project MIDAS, experienced exhausting technical difficulties. In 1963, project leaders launched MIDAS 7 and successfully used it to detect a missile launch. Project MIDAS confirmed the concept, and leaders of the Defense Support Program began launching real missile detectors in 1970. In 2000, military leaders began replacing the system with the more sophisticated Space-Based Infrared Satellites.

As leaders of foreign nations acquired wireless telephones and com-

"There is something more important than any ultimate weapon. That is the ultimate position—the position of total control over Earth that lies somewhere out in space.

This is . . . the distant future, though not so distant as we may have thought. Whoever gains that ultimate position gains control, total control, over the Earth, for purposes of tyranny or for the service of freedom."

—Senator Lyndon B. Johnson, 1958.

Photograph of launch site, Baikonur, Soviet Union, 1968. Less than one year before Americans won the Moon race, a U.S. KH-4 reconnaissance satellite, code-named CORONA, took this image showing a Soviet N1 Moon rocket at left, sitting on its launchpad.

munications satellites, military personnel launched eavesdropping satellites that could intercept messages. Military leaders launched their own communications satellites. They constructed their own weather satellite system and established navigational systems such as the increasingly commercial Global Positioning System.

Satellite Defense Surprisingly, none of these satellites can defend itself against attack from a hostile nation. If the pilot of a military jet aircraft is targeted for destruction by the operator of a surface-to-air missile, the pilot will know it and take appropriate action. Instruments in the cockpit tell the pilot that someone has locked a radar signal on the aircraft; the pilot can take evasive action or send a missile back down the radar signal to the ground. Current scenarios for ground conflict in the twenty-first century assume that adversaries will seek to destroy American satellites. The United States military depends upon its satellites to fight high-technology wars. Without its satellites, the military would be deaf and blind. Its sophisticated weapons systems would not work, and many Americans would die.

Early reconnaissance technology, 1960.
During the first years of satellite reconnaissance, spacecraft ejected film canisters that reentered the atmosphere and were retrieved by aircraft such as this C-119.

This is a startling change in the nature of warfare. During the various conflicts of the Cold War, and in regional battles that followed, no one ever disabled a satellite. Now this can be done with relative ease. Ground-based lasers can blind sensors on optical spy satellites. Technologically sophisticated countries can launch "killer" satellites that collide with their robotic foes. Nations with less money can launch buckshot into space to collide with a 17,000-mile-per-hour spy satellite and destroy it. A crude nuclear device exploded in low Earth orbit could disable the sensors on every spy satellite in the vicinity. Handheld jammers can disable global positioning signals.

To assure the survivability of their space-based assets, U.S. military leaders will deploy various space warfare devices. First, they will give their own satellites the ability to evade attacks. This will be done in a number of ways. Some satellites will swoop into different orbits if pursued. Such satellites may need to be refueled in orbit, a new technology, or carry new sources of power that permit rapid maneuvers. Scientists are working to develop small, low-cost reconnaissance satellites, easy to launch and deploy in clusters. The loss of one satellite would not cripple the system, and the loss of many would bring about quick replacement. Scientists will also develop defensive satellites that tell military leaders when a satellite is being tracked, targeted, or jammed. These defensive systems should be deployed by 2020.

Soon thereafter, the United States military will prepare itself to fight full-scale satellite wars. Military leaders will deploy lasers that can shoot down hostile satellites. They will develop "space mines." One popular concept foresees the development of "kinetic energy rods." The metal rods do not explode, but strike enemy satellites with such force that the target is destroyed.

If the military deploys lasers and kinetic energy rods, which is likely, these weapons will not be targeted at satellites alone. Space-based lasers can hit targets on the ground. They can shoot down high-flying aircraft. Kinetic rods dropped toward Earth at hypersonic speeds could destroy targets buried hundreds of feet below the ground. This sounds like science fiction, but military leaders would like to develop these systems, and the technology is within reach.

Lasers will be located in space, in aircraft, and on the ground, creating an antisatellite triad that would be difficult to destroy. Strategic thinkers hope that the existence of such a well-distributed system would deter potential enemies from striking at the U.S. satellite force, knowing that the United States could disable theirs.

The deterrence philosophy may not work. Leaders of a relatively poor nation with little reliance on satellites may be tempted to attack the U.S. satellite system because they stand to lose little in return. Satellite attacks and jamming could become the "poor soldier's" game.

All-out space warfare would contaminate the space environment. Debris from satellite destruction would likely affect other national assets such as scientific and commercial satellites. Imagine losing the Hubble Space Telescope from the fallout associated with a satellite strike. Debris from big satellites might fall uncontrollably back to Earth, creating a diplomatic backlash as it rains down on nonbelligerent nations. The use of space weapons has the potential to corrupt the physical environment long after the conclusion of any conflict, leaving a bitter legacy.

Most significant, arming the heavens violates a fundamental principle of the space age: the unimpeded operation of satellites anywhere. "Freedom of space" has been a cornerstone of U. S. space policy ever since the Eisenhower administration accepted the flight of Sputnik 1 across the United States. President Eisenhower accepted the Soviet overflight as a means of defending the right of the United States to fly satellites over the Soviet Union. "Freedom of space" is inexorably tied to U. S. space and defense policies. Military officers who deploy killer satellites may try to persuade foreign leaders that such systems are purely defensive, but those leaders are unlikely to accept such claims.

Assuring foreign leaders is hard; ensuring the protection of an antisatellite system will be arduous. Antisatellite systems are vulnerable too. Some

Reconnaissance technology. Satellites such as those in this artist's rendering (top)—advancing in sophistication from right to left—can detect events like a rocket launch from the Soviet Union, shown in a 1961 photograph (bottom), the instant they occur.

USA

officials have recommended the creation of spherical "keep-out zones" around orbiting antisatellite battle stations. Any space vehicle or space mine entering that bubble would automatically be destroyed. A set of constantly moving, bubblelike exclusion zones would violate the right of free passage in extraordinary ways.

During the latter part of the twentieth century, military leaders in the United States attempted to create a space-based antimissile defense system, popularly known as "Star Wars." One concept envisioned a large number of low-cost interceptors, called "Brilliant Pebbles," located in space. In theory, the interceptors would track and intercept ballistic missiles as they rose from enemy launchpads on their path toward the United States. Any such system created in the first part of the twenty-first century will place the interceptors and tracking radars on the ground, although the system will contain elements in space, such as early-warning satellites to detect missile launches. Though launched from Earth, missile interceptors behave like spacecraft once they leave the atmosphere, taking star sightings to adjust their trajectories. As the system expands, its advocates may renew calls for space-based interceptors.

Artist's depiction of space warfare, after 2025.
As access to space improves, governments will develop devices that protect their own military satellites and disable the satellites of others. Energy beams from this exotic spacecraft painted by Robert McCall strike enemy targets, while a swiveling pod defends the spacecraft from hostile attack.

Harbingers of Civilization During the first half of the twenty-first century, United States soldiers will return to a role they performed during the first 150 years of the republic: pioneering the American frontier. In familiar ways, military forces will preserve access to space, protect economic interests, defend military assets, and work to prevent the spread of nuclear weapons into space. They will also conduct peacetime missions on the space frontier.

This is a substantial departure from early space-age practices. At the beginning of the space age, the military did not play a pioneering role. Following the instructions of President Eisenhower, the military preoccupied itself with the development of space-based spy satellites. Eisenhower insisted that exploration activities, including human flight, be located in a newly created civil agency, the National Aeronautics and Space Administration. The U.S. Air Force recruited astronauts, but disbanded the group before the soldiers could fly. Soldiers flew on a few NASA space shuttle missions to launch military satellites and conduct experiments, a practice that ended after the Challenger accident.

The absence of armed forces in space will end soon. The first platoon of soldiers to enter space will probably travel in a military spaceplane. Since the advent of the jet airplane, military officers have sought to exploit the potential advantage gained by their ability to move troops quickly around the world. A spaceplane would enhance current airlift capabilities enormously. A military strike force would be able to reach any place in the world within an hour or two. If NASA and its corporate partners develop the spaceplane, as they plan to do, the military will surely buy a few.

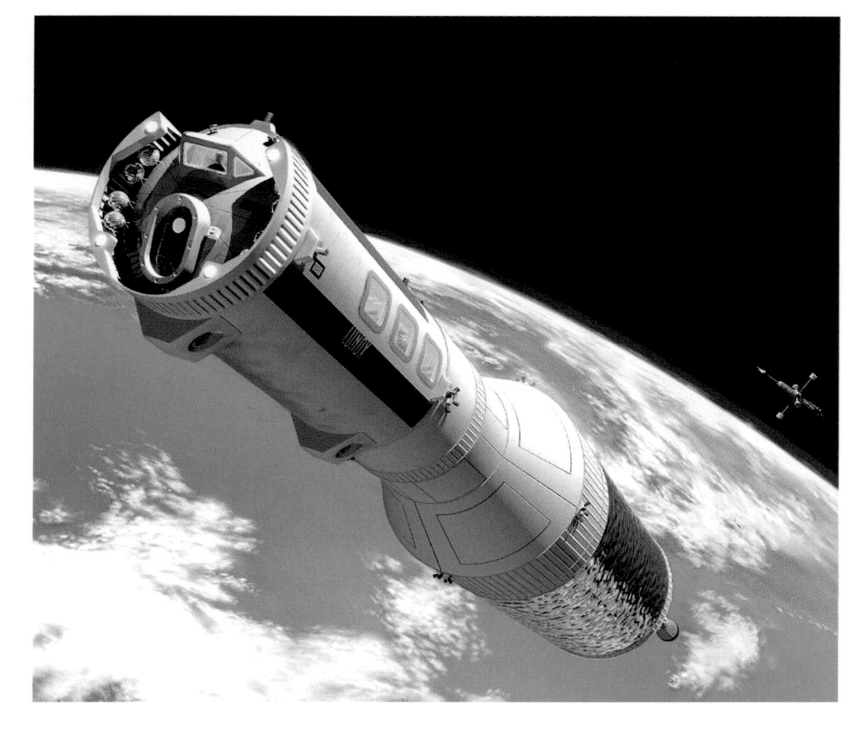

Artist's rendering of troop transport, after 2030.
As more people and commercial interests move into space, military personnel will follow. One concept envisions rocketships that can shoot soldiers to trouble spots around the globe.

Once the cost of space transportation falls, military officers will want to fly their own orbital spacecraft. They will use these shuttlelike vehicles to transport weapons to space and perform maintenance on existing satellites in the same way that NASA astronauts periodically repair the Hubble Space Telescope. Spy satellites that take pictures of Earth are similar in size and technology to space telescopes that photograph the heavens. Military forces may also use piloted spacecraft as "space bombers" from which to launch devices like kinetic rods toward targets on the ground.

As scientists and entrepreneurs spread into space, military personnel are likely to accompany them. Although the space frontier differs considerably from the American West, one aspect of the military role on the American frontier is worth remembering. For most of the time, military personnel on the American frontier performed routine activities. They raised crops, herded cattle, cut timber, quarried stone, built sawmills, strung telegraph wire, and performed the manifold duties of pioneers. They restrained lawless traders, pursued fugitives, ejected squatters, maintained order during peace negotiations, and guarded Indians who came to receive annuities. This was largely peaceful work, with the military catalyzing the processes of economic and social development. Military outposts on the frontier also served as cash markets for early settlers and as centers of exploration, community building, and cultural development.

If humans develop a base on the Moon or even an outpost on Mars, the military may perform these duties once more. Remembering the role of the U.S. Corps of Topographical Engineers and the U.S. Army Corps of Engineers in opening the American West, military leaders may propose the creation of a U.S. Corps of Space Engineers. The role they could play would be analogous to military activities in Antarctica, where the U.S. Navy maintains a contingent at the American station at McMurdo Sound and the U.S. Air Force conducts a winter resupply airdrop at the South Pole station. Similar arrangements could take place on the Moon. Military personnel could construct and maintain an isolated lunar outpost or a scientific station on the backside of the Moon. By providing logistical support, they would establish a defensive presence in space and help secure national interests.

Asteroid Strikes In 1992, a noted scientist delivered a speech before the American Astronautical Society titled "Chicken Little Was Right." The scientist claimed that humans had a greater chance of being killed by a comet or asteroid falling from the sky than dying in an airplane crash. This is true, especially as one projects the risk over a very long period of time. Mathematical calculations confirm that every person alive today faces 1 chance in 5,000 that he or she will be killed by some type of extraterrestrial impact during his or her lifetime. Several thousand meteorites, comets, and asteroids cross Earth's orbit, and many small pieces enter the atmosphere every day. One need only look at the craters on the Moon to verify that celestial bodies make fine targets for comets and asteroids. Within time, a really big one will hit Earth with disastrous consequences. The scientist urged the government to catalogue all Earth-crossing asteroids, track their trajectories, and develop countermeasures to destroy or deflect objects on a collision course with Earth.

Most of the people who heard the speech departed with mixed feelings. They recognized the reality of what they had been told and also denied that it bore any relationship to their lives. The audience knew that a great galactic asteroid might have killed the dinosaurs, a popular theory. That event, however, occurred 65 million years ago, not exactly a current threat. The United States had just achieved a great triumph, defeat of the Soviet Union in the Cold War, and stood omnipotent as the world's remaining superpower. Now another threat was looming: this time it was from space, and there was no technology to defeat it.

Public perspectives changed in 1994. In July of that year, the comet Shoemaker-Levy 9 smashed into Jupiter. The Hubble Space Telescope recorded the event in frightening detail. The comet, originally 6 miles wide, broke into pieces as it fell through Jupiter's gravitational field. Hitting Jupiter, the largest fragments left shock waves as wide as the diameter of Earth. The destructive power of the impact made believers of many skeptics. Had Shoemaker-Levy 9 hit Earth, the results would have been catastrophic.

Throughout history, asteroids and comets have struck Earth, leaving impact craters that are visible from space. The asteroid that allegedly killed the dinosaurs left a crater 186 miles wide on the Yucatán peninsula in Mexico. The object was probably 6 to 9 miles wide. Due to erosion and vegetation, the impact zone is not discernible to a casual observer on the ground, but the view from space is clear.

At some point in the future, an asteroid or comet will target Earth and threaten lives. It is as inevitable as the rising of the Sun. Computer simulations reveal that waves spawned by a 3-mile-wide object dropping into the Atlantic Ocean would overwhelm the East Coast of the United States and reach the Appalachian Mountains. New York City would remain underwater for hours.

"The Air Force and other government agencies, industry and the citizenry of America are beginning to clearly see the potential represented by space. . . . In the decades to come, spacepower will accomplish many of the same functions that airpower accomplishes today.

Spacepower will encompass space superiority, space control, space surveillance missions, information superiority, and the list goes on. I envision a day when spacepower will also represent the ultimate in rapid global mobility and global precision attack."

—Gen. Howell M. Estes III, USAF, 1997.

Experts disagree on how best to respond to these threats. Most scientists want to wait until a specific threat is identified before developing countermeasures. Military leaders argue for a more aggressive approach. They want to develop technologies by which the United States as the world's leading space power could deflect or destroy an Earth-threatening body.

Scientists have organized major sky surveys. From the University of Arizona's Steward Observatory, on Kitt Peak, a robotic telescope systematically photographs the sky, while a computer analyzes the images and alerts astronomers when moving objects appear. The Spacewatch system routinely detects two thousand asteroids per month, of which about three turn out to be near-Earth objects. NASA and the U.S. Air Force operate a Near-Earth Asteroid Tracking telescope at Haleakala, on the island of Maui, which uses a similar technology.

Within a decade, scientists will possess an inventory of all asteroids of substantial size in the solar system. Tom Gehrels, director of Spacewatch, prom-

ises that "if there is an asteroid out there with our name on it, we should know by about the year 2008."

Military leaders are planning defensive strategies. Deflection is a viable option for objects discovered years in advance of the likely collision. With much advance notice, an asteroid would need only a small nudge to miss Earth. A fraction of a single mile per hour would be fine. This could be accomplished in a number of ways. Scientists could crash a rocket into one side of the asteroid. Such a strategy would be most effective for asteroids smaller than 300 feet in diameter. They could detonate a neutron bomb in space some distance from the asteroid. The explosion would heat one side of the asteroid, causing chips of rock to explode away. By the principle of action and reaction, the asteroid would drift slowly in the opposite direction. Alternatively, engineers could build a new type of rocket engine and attach it to the surface of the asteroid. Operated for several years, the engine would nudge the asteroid into a new orbit.

Artistic concepts of meteor strikes.
The greatest threat to world security may come not from Earth, but from beyond it. A 1949 Chesley Bonestell painting illustrates how a 150-foot-wide object would devastate Manhattan Island, New York (left). A larger asteroid or comet may have ended the reign of the dinosaurs. Space-based images suggest that a 6- to 9-mile-wide object struck Mexico's Yucatán peninsula 65 million years ago, as seen in this artist's rendering (right).

What about a really big, civilization-killing asteroid? Military might, coupled with science and technology, could prevent such a catastrophe. A rocket full of explosives (probably nuclear) could be launched toward the object. An explosion on the object would not be wise since this might divide it into equally destructive parts carrying the force of multiple warheads. Instead, the explosion would occur in the space adjacent to the object in such a way as to deflect its course.

Given current technology, scientists would need to detect large asteroids years, even decades, in advance for preventative measures to be effective. Lead times would have to increase for big asteroids on close trajectories. No one knows for sure whether the military could meet this challenge. Deflection of an Earth-bound, killer asteroid, in the words of the late Congressman George Brown, "will be one of the most important accomplishments in all of human history."

08 : The Greening of Space

* * * * * *

preceding pages

New York City, vision and reality.

Prior to the space age, artists could only imagine how the Earth looked from above the atmosphere. A 1983 Landsat 4 image (right) confirms Chesley Bonestell's presatellite 1949 vision of the same view (left).

The twenty-first century may prove to be the most difficult for humanity since the advent of the Renaissance. During the century earthlings will face three great environmental challenges: overpopulation, resource depletion (especially fossil fuels), and environmental degradation. Without space-based resources—especially remote sensing satellites that monitor Earth—humans will not be able to control these trends. Civilization could slip into a new dark age, made more terrifying by remaining nuclear weapons and toxic waste.

Humans can use space as a place from which to monitor the health of Earth, maximize natural resources, and spot polluters. By joining space with activities on the ground, humans have a fighting chance to protect the environment in which they live. Using space to protect Earth will be as important to twenty-first-century history as Moon landings were to the twentieth.

At the same time, humans will confront the consequences of environmental degradation in space. Orbital debris, derelict spacecraft, and satellites reentering the atmosphere have already created hazards around Earth. Proposals to strip-mine the Moon and asteroids make many people blanch; how dare humanity, having fouled Earth, destroy the pristine quality of extraterrestrial bodies? Aeronautical engineers want to develop nuclear-powered spacecraft for planetary exploration; environmentalists have fought this trend. The environmental movement will move into space.

Equally significant is the rising concern about biological contamination. Scientists take special precautions when launching spacecraft to places like Mars to ensure that the spacecraft do not carry microorganisms to those bodies. As samples return, public concern about reverse contamination will grow. Bringing extraterrestrial samples to Earth presents a hazard for this planet; what if some totally unknown organism attacks life here?

All of these issues—the use of space for monitoring Earth, environmental degradation in space, and biological contamination—promise to create a new perspective on space exploration. As a result, humans in the twenty-first century will witness the greening of space.

The Club of Rome Sounds the Alarm Never in the course of history has humankind faced so much danger. So indicate members of the Club of Rome, an international organization formed in 1968 to investigate global change. In their classic study, titled "The Limits to Growth," club researchers explain how five fundamental factors affect the future of humankind. Armed with computer-based simulation techniques, researchers investigated the interaction of rapid population growth, accelerating industrialization, depletion of natural resources, increasing pollution, and agricultural productivity.

For a while, rapid industrialization and increased agricultural productivity outpace pollution and population growth. The worldwide standard of living remains high. At some point, however, Earth can no longer support a large, industrial civilization. In all of the computer models, a terrible clash occurs. Population, industrialization, and food production enter a drastic and uncontrollable decline. According to the computer models, Earth reaches this point of no return during the twenty-first century.

Can humans alter these trends and balance economic growth with ecological stability? Optimists insist that technology can stave off eventual doom. Advocates of space exploration believe that this technology will come from space. Their prognostications are both misleading and wise.

following pages
Environmental degradation, photographs from space, 1985–1994.
Space-based technologies reveal precisely how human activity affects the Earth's environment. Left to right: Air pollution obstructs the view of Mexico City; Kuwaiti oil wells burn in the aftermath of the Gulf War; deforestation reduces the Brazilian rain forest; and the burning of fossil fuels darkens the snow-covered landscape around the Russian town of Troitsk.

Gaia: The Earth as Living Organism Before humans can save the biosphere, they must understand how it works. In the 1970s, James E. Lovelock, a British scientist, formulated a controversial biochemical theory while working on a NASA-sponsored study. Seeking a means for detecting life on Mars, Lovelock discovered some strange properties of Earth. He called his findings the Gaia Hypothesis, a provocative label drawn from the name of the Greek Earth goddess. According to Lovelock, Earth as a whole is a dynamic organism with features similar to those found in a living being. Lovelock and his followers believe that Earth's living organisms and inorganic material interact in such a way as to reshape the biosphere. They change the atmosphere and regulate surface temperatures, producing conditions conducive to the maintenance of life. Vital organs like tropical rain forests keep the planet clean; greenhouse gases maintain a comfortable climate. When one element of the biosphere moves out of its normal range, other elements bring it back into line. Earth has operated like this for millions of years, until humans came along.

Humans have released greenhouse gases in unusual quantities. When the human body senses a viral attack, it elevates its internal temperature as a way of killing the invader. So apparently does Earth. Although scientists disagree on the magnitude of the problem, many predict that average global temperatures will rise 3 to 8 degrees Fahrenheit in the twenty-first century. This will appreciably alter patterns of rainfall and snowfall. Local climates will change. Many plants and animals will perish, and human civilization will face serious disruption. As ocean waters warm and expand and ice caps melt, sea levels could rise by 12 inches or more, flooding low-lying islands and endangering coastal areas where millions of people live.

According to the Gaia Hypothesis, as elaborated by American biologist Lynn Margulis, evolution of living beings is not driven by chance mutation

"Some of civilization's more unfortunate effects on the environment are . . . evident from orbit. Oil slicks glisten on the surface of the Persian Gulf, patches of pollution-damaged trees dot the forests of central Europe. Some cities look out of focus, and their colors muted, when viewed through a pollutant haze. Not surprisingly, the effects are more noticeable now than they were a decade ago. An astronaut who has flown on both Skylab and the space shuttle reported that the horizon didn't seem quite as sharp, or the colors quite as bright, in 1983 as they had in 1973 . . ."

—Astronaut Sally Ride, May 1986.

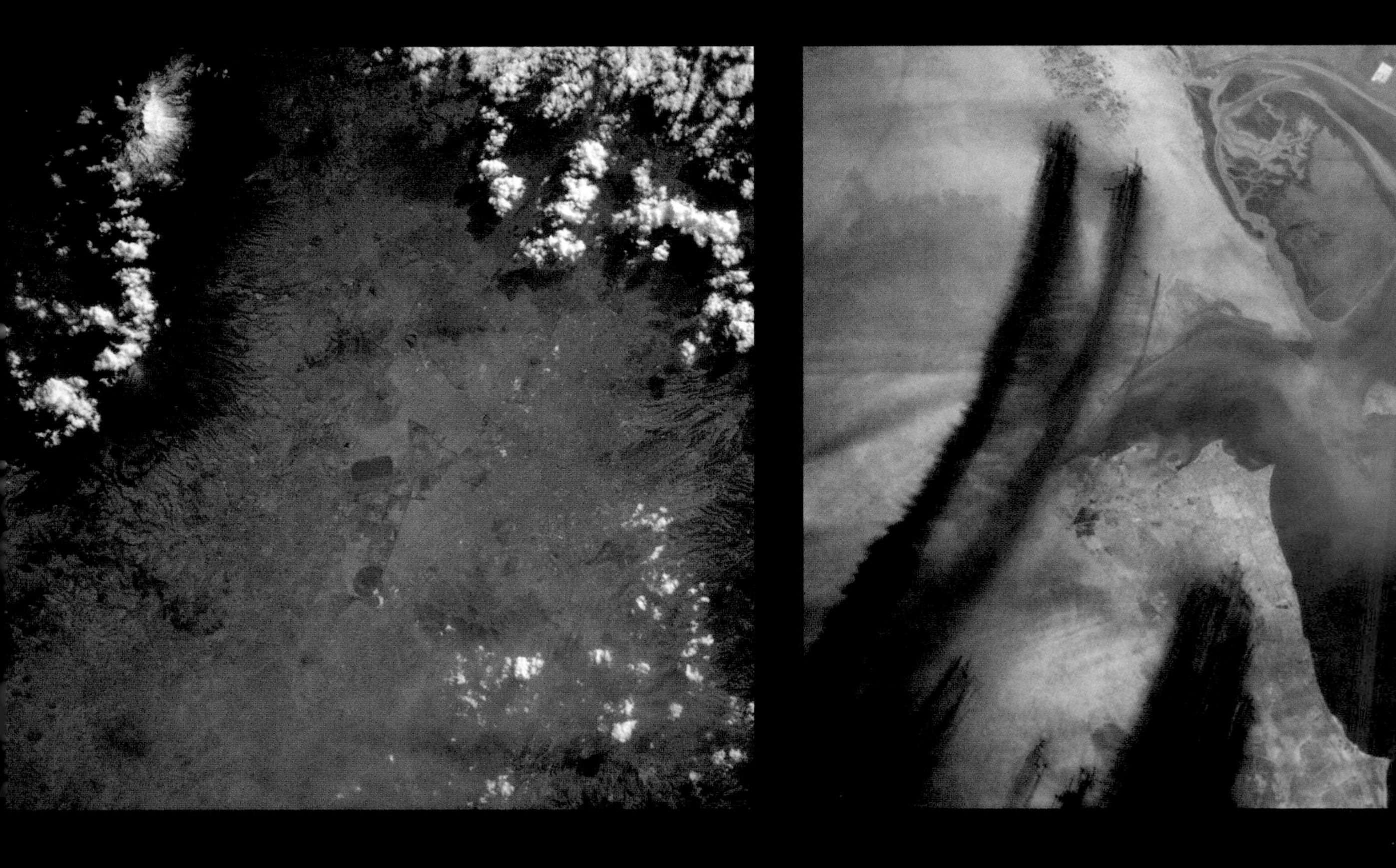

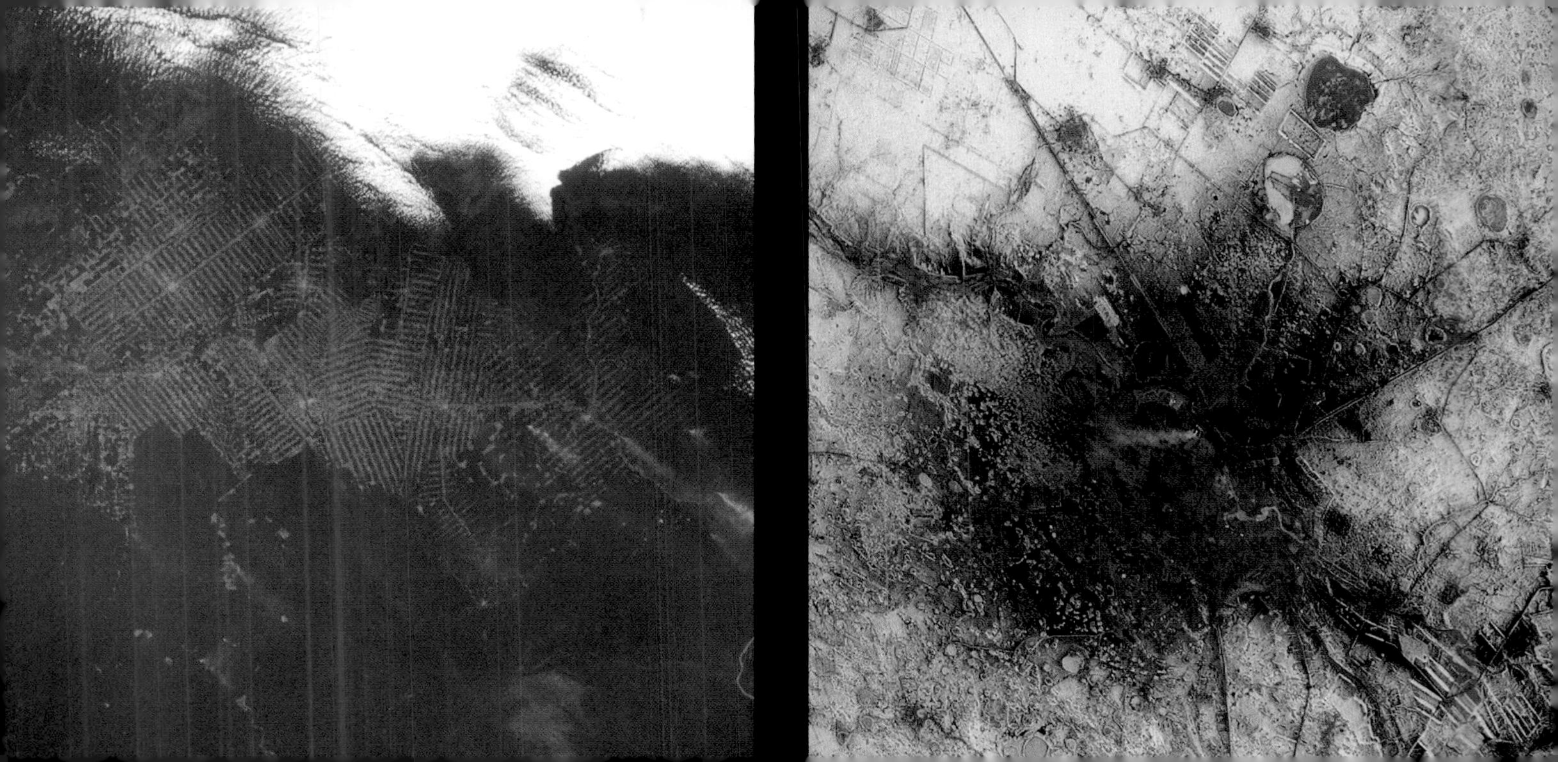

and competition between species. Rather, the symbiosis between organic and inorganic matter inherent in the Earth's biosphere drives evolution. Cooperation between organisms and the environment are the chief agents of natural selection—not competition. The fundamental components of Gaia theory will inform humans as they try to keep Earth green during the twenty-first century. The planet is a "superorganism," and cooperation between elements of the biosphere determines the state of its being.

Artist's concept of a space colony, 1975.
During the 1970s, physicist Gerard K. O'Neill proposed the establishment of very large colonies in the emptiness of space as a means of relieving population pressure on Earth. The idea, which attracted many followers, is not a feasible solution in the short term.

Space Colonization To a number of visionaries, space exploration offers a special solution to global stress. Visionaries like Carl Sagan and Gerard K. O'Neill talked about moving humans off Earth. Such migration, they argued, would reduce pollution and preserve natural resources.

O'Neill, a Princeton University physics professor, published detailed plans for artificially constructed space colonies during the 1970s. O'Neill urged humans to establish colonies in very large, rotating spacecraft placed at gravitationally stable points throughout the solar system. Colonists would live in clean, climate-controlled environments, with trees and lakes and blue skies spotted with clouds inside each colony's rim. Animals and plants endangered on Earth would thrive on these cosmic arks; insect pests would be left behind. Solar power, directed into each colony by huge mirrors, would provide a constant source of nonpolluting energy.

O'Neill believed that the first colony could be completed by 2005. Each fully developed colony would provide room for 10 million humans, plus desirable plants and animals. Emigration to newly constructed colonies, O'Neill estimated, would reverse the population rise on Earth by 2050.

Many sensible people flocked to O'Neill's schemes. Supporters founded the L-5 Society, named for one of the location points, as a means of arousing "public enthusiasm for space colonization." The U.S. Congress conducted hearings, and NASA officials commissioned feasibility studies. Interest in space colonization spawned imitation, including elaborate proposals for subjecting Mars to terraforming, a process by which otherwise uninhabitable spheres are transformed into livable bodies with Earth-like atmospheres.

Alas, these dreams will not come to pass soon. Such fantastic schemes contribute little to the solutions required to sustain the planet. They constitute a form of denial in that they direct public attention to solutions that are technically infeasible. Humans will learn to live on the harsher portions of their own planet—such as polar regions or under the seas—before they move in large numbers to space colonies or Mars. As a means of promoting resettlement, government officials might construct mirrors in space that beam sunlight to polar regions otherwise cloaked in winter darkness. Humans had best learn to take care of their own planet before abandoning their home for starry evacuation schemes.

The Solar Power Solution Since the burning of fossil fuels precipitates global warming, space enthusiasts have offered replacement strategies. One of the most compelling schemes involves the construction of solar power satellites that direct energy beams toward Earth. The last one hundred years have been called the carbon century, a reference to the fossil fuels that energize the world. Space advocates believe that people in the twenty-first century can inaugurate a period of limitless growth based on the power of the Sun.

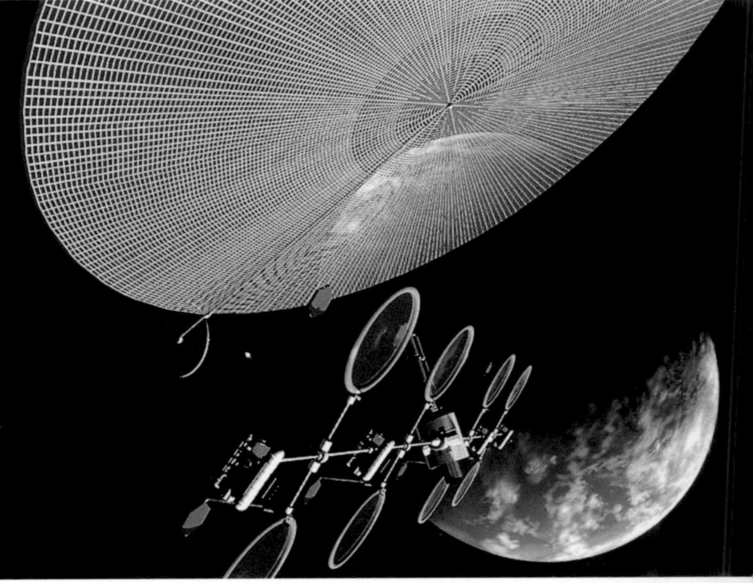

Artistic renderings of solar-powered satellites, after 2035. Advocates of one persistent vision foresee the use of solar power to replace fossil fuels. A SunTower (left) or SolarDisk (right) could direct high-energy beams to power stations on Earth. Solar power, however, is not likely to be economically competitive compared with twenty-first-century ground-based alternatives like hydrogen-powered fuel cells.

Outer space is full of solar power. It is clean, renewable, and more plentiful than any other energy source on Earth. With this in mind, advocates have recommended the construction of solar power stations orbiting Earth and stationed on the Moon. Such stations would collect solar energy in large mirrors and beam it to ground stations on Earth, where it would be converted into electricity. This is an appealing idea. It would seem to resolve one of the thorniest problems identified by the Club of Rome: finding clean power to fuel the engines of the world. Like space colonization, however, it is not a workable concept.

Why is this so? First, the cost of generating electric power from space is not competitive with Earth-based sources. In 1999, people on Earth generated nearly 12 billion kilowatt-hours of electrical energy from all sources, including petroleum, natural gas, coal, hydroelectric dams, and nuclear power plants. Energy use grows at a rate of about 2.4 percent per year. Conversion of even a small percentage to space-based production would require an investment that no political leader is likely to propose. The cost would far exceed the price of electricity generated by substitute fuels.

Second, solar power sources can be hazardous. Any object crossing a beam of energy from space would instantly be destroyed–flocks of birds or even the stray airplane. People worried about the potential danger of electromagnetic emissions from cellular telephones or high-tension lines are naturally concerned about the effects of high-energy beams on people nearby.

Third, a space-based solar power system of adequate size is well beyond human technological capabilities for the foreseeable future. One concept, studied by NASA and the Department of Energy in 1979, envisioned the construction of large solar power platforms in geostationary orbit. Each platform would deliver 5 gigawatts of electricity through a microwave beam to a ground station on Earth. Sixty such platforms could provide about 2.5 percent of Earth's energy needs. These are very large structures. They would need to be assembled by hundreds of astronauts using materials delivered by rockets that do not exist. The platforms would require the development of more than one hundred new, high-risk technologies. To start work on such a system would cost twice as much as the Apollo expeditions to the Moon, nearly $300 billion in inflation-adjusted dollars. Not surprisingly, government leaders have not funded the idea.

Solar power is not an entirely senseless idea. Some experimentation will take place. NASA engineers have investigated two concepts that might provide moderate amounts of low-cost solar power. One of these, called SunTower, consists of a modular, self-assembling satellite that would operate in an orbit relatively close to Earth. It arranges a series of solar collectors along a very tall pole, so tall that gravity helps stabilize its position in space. A second concept, SolarDisk, would be placed in geostationary orbit. This system is also modular, but larger, and could be assembled through onboard robotics. NASA officials would like to test both concepts by 2035.

The feasibility of these concepts remains uncertain. The technologies needed to perfect space-based solar power are neither easy to develop nor readily available. Scientists and engineers may require as much as a century to deliver a workable system. If so, it will not come on-line in time.

Earth-monitoring photographs taken from space, 1984–1991. Shuttle astronauts photographed smoke plumes from fires used to clear Amazon rain forests in the state of Rondônia in western Brazil (top left). Studying the spectral and spatial characteristics of auroral emissions, they took pictures of the aurora australis, or southern lights (top right). Space-based observations have also improved the understanding of weather dynamics. The 1985 Hurricane Elena, with winds in excess of 110 miles per hour, was photographed as it prepared to strike land near Gulfport, Mississippi (bottom). Scientists will expand their use of space as a means to monitor Earth's natural environment as the twenty-first century proceeds.

For the short term, hydrogen is more likely to provide a substitute for greenhouse gas–producing fuels. The next one hundred years may be the hydrogen century. Hydrogen is abundant on Earth, being one of the principal elements in water. It can be used to power fuel cells, devices that work like batteries. Burned with oxygen, hydrogen powers rocket engines, such as those on NASA's space shuttle. Fuel cells and hydrogen-oxygen engines, spin-offs from the space program, are rapidly becoming cost-competitive. In the same way that carbon fuels fired the industrial engines of the twentieth century, hydrogen is expected to power the twenty-first.

In the beginning, humans will obtain hydrogen from fossil fuels. As the century progresses, the cost of extracting hydrogen from ordinary water may fall. Extraction from water requires electrical energy. Some of that electricity could be generated from solar panels on the surface of Earth or from a combination of ground panels and solar satellites.

Monitoring Earth's Health The most serious use of space for environmental protection involves satellite monitoring. Beginning in the 1980s, most Americans became aware of the twin planetary symptoms of human abundance–global warming and ozone depletion. Human activities, primarily the burning of fossil fuels and the removal of land cover, have increased atmospheric concentrations of greenhouse gases, thereby decreasing the amount of heat expelled to space. Scientists expect this to increase surface temperatures, alter cloud cover, change precipitation patterns, and increase the likelihood of hurricanes, floods, and droughts.

Debate over the greenhouse effect is politically charged. People committed to fossil fuels dispute the existence of a problem. They argue that compensating factors, such as increased cloud cover, will counteract warming effects. Unfortunately, scientists do not know exactly how the elements of the biosphere interact, and they lack devices that can measure the effects of induced change.

Shortly after the space age began, NASA officials began an effort to collect this sort of information. They pioneered a new discipline, called Earth system science. Using a variety of instruments, Earth system scientists study land cover, agricultural productivity, oceans, atmospheric changes, climate, ozone depletion, and natural hazards. They seek knowledge that can provide a scientific basis for global intervention.

Space offers a unique vantage point for collecting information about Earth's land, atmosphere, ice cover, oceans, and biota that is unobtainable any other way. Useful information has been collected through Earth resource satellites already launched. Landsat 1, launched in 1972, photographed land-use patterns and changed conventional views of the planet. Imaging capabilities

improved with the 1999 launch of Landsat 7. The newest satellite maps all surface areas on Earth every sixteen days, providing the means for monitoring land-use changes that affect global change.

These capabilities will expand dramatically during the first half of the twenty-first century as additional satellites are launched. NASA will initiate a number of efforts aimed at collecting data about the planet, known collectively as the Earth Observing System. Satellites will survey atmospheric physics and chemistry, weather and climate change, land surface alteration, terrestrial ecosystems, oceanography, and marine ecosystems. Scientists will collect data from a new series of Geostationary Operational Environmental Satellites, routinely used in the past to photograph weather conditions over a fixed portion of the globe. They will use Polar Operational Environmental Satellites to investigate numerous atmospheric and surface features and forecast climatic change.

NASA scientists are also employing the Tropical Rainfall Measuring Mission. They will develop smaller satellites through the Earth System Science Pathfinder mission and test new monitoring techniques through the New Millennium program. Scientists hope to launch a New Millennium Earth observing satellite every eighteen months. By 2020, a fleet of Earth resource-monitoring satellites will be in orbit, providing continuous data on the planet's land, sea, and atmosphere. Space-based data, combined with information gathered from balloons and airplanes, should equip political leaders with the knowledge required to dispel myths and make sound recommendations about global warming. The history of ozone depletion shows how this may occur.

"What makes it possible to dream audaciously of understanding the web of global Earth processes is remote sensing . . . Through simultaneous measurements made of the surface, atmosphere, and oceans, we can begin to see how these systems are interrelated. The hope is that this effort will be the EKG and stethoscope for the planet."

—John F. Mustard, Professor of Geological Sciences, 1989.

Filling the Ozone Hole Like global warming, ozone depletion also excites environmental concern. Most Americans are now aware that the Earth's ozone layer protects all life from the Sun's harmful radiation and that human activities have damaged this shield. Atmospheric ozone is concentrated in the stratosphere, about 12 to 25 miles above the Earth's surface. Ozone is a molecule containing three oxygen atoms. It is blue in color and has a strong odor. Normal oxygen, which living creatures breathe, has two oxygen atoms and is colorless and odorless. Ozone is much less common than normal oxygen. For every 2 million molecules of oxygen in the atmosphere, only three molecules of ozone can be found.

While the amount of ozone in the atmosphere is quite small, it plays a key role in protecting life. Most important, it absorbs the portion of ultraviolet light called UVB (ultraviolet B light). UVB has been linked to many harmful effects, including skin cancer and cataracts, and damage to crops, surface-dwelling animals, and marine life.

Ozone concentrations vary naturally with sunspots, seasons, and latitude. These processes are well understood and predictable. Having studied ozone levels over several decades, scientists understand how ozone levels behave. Historically, each natural reduction in ozone is followed by a period of recovery. Convincing scientific evidence collected during the twentieth century revealed that recovery cycles were falling below expected norms.

One consequence of ozone depletion is the appearance of an ozone "hole" over Antarctica. This has occurred annually during the Antarctic spring since the early 1980s. The so-called hole is in fact a large area of the stratosphere with abnormally low amounts of ozone. Ozone levels in the hole fell more than 60 percent during the worst years. Research revealed that ozone depletion also occurred over latitudes that include North America, Europe, Asia, Africa, Australia, and South America. Over the United States, ozone levels fell 5 to 10 percent depending on the season. Accordingly, ozone depletion is a global issue and not just a problem at the South Pole.

Scientific investigation provided documented evidence linking chlorofluorocarbons (CFCs) to the depleted ozone layer. These incontrovertible findings led to a series of international agreements that limited the use of CFCs. In 1985, representatives attending the Vienna Convention for the Protection of the Ozone Layer adopted a plan that formalized international cooperation on this issue. Additional efforts resulted in the signing of the Montreal Protocol on Substances that Deplete the Ozone Layer in 1987. Because of measures taken under the protocol, emissions of ozone-depleting substances began to fall. Assuming continued compliance, stratospheric levels will peak in a few years and then slowly return to normal. Natural ozone production should heal the ozone layer in about fifty years. This is good news.

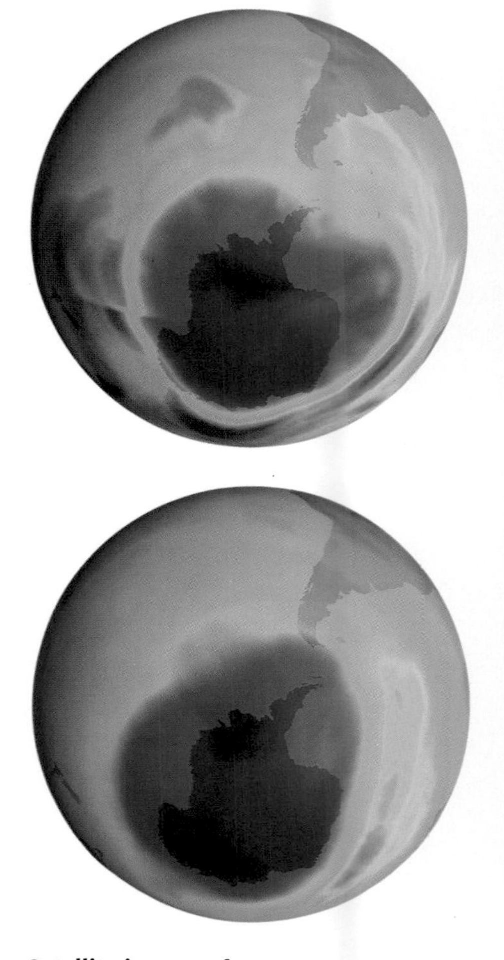

Satellite images of ozone depletion.
Placed onboard a succession of spacecraft, Total Ozone Mapping Spectrometers provide daily images of the ozone hole of Antarctica. From 1998 (top) to 2000 (bottom), the hole expanded northward toward South America.

Each year, a collection of scientific organizations, including NASA, measures the chemical composition of stratospheric clouds that trigger ozone loss over the South Pole. In 1999–2000, 350 researchers took part in the largest cooperative field investigation of polar ozone, using satellites, airplanes, balloons, and ground instruments to study the hole. They sampled clouds for evidence of CFCs. They studied strange phenomena like the "cool pool" of air that exists naturally in the stratosphere and air waves caused by wind flowing over mountain ranges.

Through observations such as these, scientists are coming to understand the complicated chemistry associated with ozone loss. With this knowledge, they locate causes and recommend cures. Policy makers can then negotiate and sign international agreements to resolve the problem and champion further monitoring of restoration efforts.

Regulating Human Behavior from Space As the ability to monitor Earth improves, the ability of humans to hide environmental harm declines. Sociologist

Satellite photographs of climate change.
Images taken by satellite revealed the intense effects of El Niño on the ocean's ecosystem. El Niño appeared in 1997, indicated by the warmer than normal temperatures shown in red (left). Plankton, shown in green, largely disappeared (center). Two years later, the cool waters associated with La Niña raised nutrients from below the surface and enabled an enormous bloom of plankton, an important rung on the oceanic food chain (right).

Amitai Etzioni has expressed concerns about what he calls the "death of privacy"—initially a reference to the practice of mining the Internet for personal information that corporations can use. The "death of privacy" is a legitimate and timely concern, one that will increasingly involve space scientists. Scientists plan to use satellite imagery to identify activities that harm the environment. While most experts view this as a necessary development for maintaining the biosphere, it involves a further loss of privacy.

A foreshadowing of what might take place occurred in the last decade of the twentieth century. A film shot by astronauts on the NASA space shuttle strikingly revealed the widespread burning of Amazon rain forests to make room for human activity. Smoke clouds were readily visible from space. The images excited people concerned with environmental destruction and sparked international pressure to restrict, if not to halt, the practice.

During the twenty-first century, images from space will provide extensive information for government regulation. The information will identify polluters and help build cases for their restraint. Images from space will verify compliance with environmental agreements in much the same manner that military satellites confirm adherence to arms control treaties.

Much of this will be accomplished through nanosatellites. These are very small sensors, some no larger than three-layer birthday cakes. Engineers are designing nanosatellites that can be mass-produced at low cost and launched together in bunches. Various constellations of fifty to one hundred satellites flying in formation around the planet will provide detailed information on all aspects of the Earth's biosphere, both natural and human-induced. Some will measure natural phenomena like the solar wind from multiple points in space. Others will monitor human activities on Earth. In countries where weak monitoring provides a shield for polluters, nanosatellite observations can protect the environment, but at a cost to privacy.

Protecting Outer Space As machines in space protect Earth, who will protect space from the machines? Humans have already fouled the extraterrestrial realm. The Apollo landing sites on the Moon resemble unkempt campsites on Earth, with human trash, discarded artifacts, and disturbed natural landscapes. Lunar explorers did not adhere to the prime directive of naturalists: leave nothing but footprints, take nothing but memories.

Space advocates often evoke the American frontier as a happy metaphor for space exploration. In doing so, they typically ignore the mining spoils, clear-cut forests, destroyed fisheries, and nuclear waste sites that frontier practices produced. Historian Patricia Nelson Limerick has examined the relationship between space exploration and frontier ethics and urged space advocates not to repeat these practices again. In her mind, frontier settlement denotes conquest of place and peoples, exploitation without environmental concern, wasted natural resources, political corruption, shoddy construction, brutal labor relations, and financial mismanagement. Limerick suggests that NASA officials protest all efforts to portray their agency as advancing frontiers. To someone familiar with American history, Limerick says, such a reference is an insult to the honor of the organization.

Artist's depiction of nanosatellites, projected for 2003. The ability of scientists to closely monitor the Earth-space environment requires the development of small, low-cost satellites that can operate concurrently in large constellations. During the first decade of the twenty-first century, NASA scientists plan to test miniature satellites that are not much larger than birthday cakes.

Environmental concerns about space exploration will expand in the twenty-first century. Natural reserves will be established on the Moon. Environmental regulations for private companies operating in space are inevitable. Some private companies will make a profit cleaning up space debris. Scientists want to collect samples from other bodies in the solar system and return those samples to Earth. This offers an exciting scientific opportunity. It also involves potential risks that need to be well managed.

As increasing numbers of spacecraft leave Earth, especially for Mars, scientists may find themselves caught between the Scylla of discovering life and the Charybdis of having to deal with it. Explorers could contaminate extraterrestrial life in much the same way that clumsy archaeologists have disturbed historic sites on Earth. Scientists bringing samples back to Earth will face public fears about an "Andromeda Strain," a reference to the Michael Crichton novel in which a microorganism retrieved from space produces violent human death.

People who work on space projects take special precautions to guard against such possibilities. The first astronauts to return from the Moon were placed in quarantine to ensure that they did not harbor some strange microorganism. The lunar samples they brought back were likewise isolated, even though scientists knew that the Moon harbored no life. Scientists routinely sterilize spacecraft bound for Mars. Engineers baked the Viking landers, the first spacecraft to search for life on Mars, for forty hours at temperatures reaching 234 degrees Fahrenheit. Sterilization strains spacecraft components

Precious Earth, 1989.
Following the release of the spacecraft Magellan, headed to Venus on a radar mapping mission, shuttle crew members photographed Magellan's destination in the sunset glow above Earth. Space travel allows humans to see Earth as it really is—a small, precious globe with only a fragile band of atmosphere to protect it from the assaults of deep space.

and increases the odds of failure, but it must be done.

Politicians will insist that scientists take strict precautions to ensure that extraterrestrial samples brought to Earth are carefully contained. Containment must be verified during the entire return trip and maintained through reentry of the spacecraft, as well as during transfer to receiving facilities. Once in storage, continual monitoring must ensure that no outward contamination takes place. Translating such policies into workable practices will be hard. Success cannot be guaranteed, and the potential risks resulting from error are substantial.

Strict containment of a different sort is also required. Extraterrestrial materials must be protected from earthly contamination. No scientist wants to discover that the life on an extraterrestrial sample crawled in through an unsecured lab door. Scientists want to protect the pristine and unaltered state of extraterrestrial bodies. Viewed simply, the need for preservation is driven by two distinctly different concerns: preserving extraterrestrial life and protecting Earth. The former emphasizes keeping materials in, while the latter emphasizes keeping contaminants out.

Extraterrestrial sample return missions should generate considerable public interest. Excitement about the discovery of potential extraterrestrial life will be sobered by concern about adverse effects. The success of sample return missions will depend, in part, on how scientists address protection issues.

The challenges of the twenty-first century are intimidating. Limited natural resources, heightened demand for them, and pollution of the planet all require global attention. Conservation issues no longer affect Earth alone; with space technology, careless humans can abuse resources everywhere. Space-based technologies provide many of the tools needed to combat environmental degradation. In utilizing these tools, humanity can begin the "greening of space," preserving the home planet and the solar system as well.

"Feats such as going to the Moon, orbiting the earth for weeks at a time, or installing and repairing the instruments that expand our knowledge, are all celebrations of everything we Americans are supposed to be. When we decide to do so, we solve problems. We figure things out. We go into space.

I know such concepts as a permanently manned orbiting science station and other NASA programs are not as glamorous as going to the Moon. And Lord knows that our problems here on our world need our attention, resolve and service.

But to choose not to go into space, to decide that our days of discovery and conquest there are over, to cease or curtail funding for the one American program that exists solely to advance the horizon for all mankind would be, I think, equal to limiting the grand power of pure inspiration, hampering our manifest destiny and taking away the best part of all of us."

—Tom Hanks, actor, 1995.

09 : Understanding the Universe

* * * * * *

preceding pages

Artistic concepts of Saturn as seen from Titan.

Chesley Bonestell painted a stunning illustration for a 1944 issue of *Life* magazine (left). The atmosphere on Titan, one of Saturn's moons, which the artist guessed to be clear blue, is in fact opaque. In 2004, scientists hope to drop the Huygens probe—part of the Cassini mission to Saturn—into Titan's thick methane atmosphere (right).

The telescope—the primary instrument for understanding the universe–is changing in unimaginable ways. Four hundred years ago, a Dutch optician, Hans Lippershey, aligned two pieces of glass in a manner so as to focus a magnified image on an observer's eye. Galileo Galilei heard of the instrument and in 1609 produced a telescope through which he observed the moons of Jupiter and the phases of Venus, substantiating the Copernican theory that all of the planets revolve around the Sun.

In the four centuries that followed, astronomers learned how to construct larger telescopes. They placed them on mountaintops, better suited for observing cosmic details. They learned how to fly telescopes in space, above the obstructing interference of the Earth's atmosphere. For most telescopes, however, the basic technology remained the same–whole pieces of glass that when aligned magnified light waves from distant objects. All that is about to change.

The Telescope Revolution In 1990, astronauts on the space shuttle Discovery deployed the Hubble Space Telescope, the first of four great observatories that NASA officials planned to place in space. After an important correction to its primary mirror, the telescope produced images of astonishing clarity. It recorded images of stellar nurseries, planetary nebulae, exploding stars, and the event horizons of black holes. Images tantalized scientists, inspiring them to desire larger and better telescopes. Space telescopes, however, cannot grow indefinitely. To support the 2.4-meter-diameter mirror in the Hubble Space Telescope, scientists produced an instrument that weighs 25,500 pounds. Scientists want to install an 8-meter-diameter mirror in the Next Generation Space Telescope, Hubble's successor, and even larger ones to follow. Built along the same lines, larger mirrors require orbiting observatories of impracticable size.

The solution to this problem is conceptually simple but technically challenging. Astronomers must install larger mirrors in smaller instruments. On its face, this seems impossible. Given traditional telescope technology, bigger cannot be smaller. Scientists will solve this apparent contradiction by producing telescopes that do not depend upon large, rigid mirrors. Instead, scientists plan to collect images on divided pieces of glass that move. They are already doing so, beginning with telescopes on the ground and moving through space-based observatories.

Telescopes of the twenty-first century will consist of smaller and lighter pieces of glass whose exact positions are computer controlled. In some cases, computers will reshape whole pieces of flexible glass. In other cases, divided pieces of glass will be joined together like a jigsaw puzzle to form large mirrors. By dividing glass, scientists can squeeze large telescopes inside small spacecraft. In one concept, scientists would fold the mirrors. Once the instrument reached its proper orbit, the mirrors would unfold like a flower bud to create a new window on the universe.

Many small pieces of glass need not be joined together. The pieces could have gaps between them, further reducing the weight of the overall instrument. Four relatively small telescopes, mounted on a 300-foot-long truss, would produce images forty times more precise than those captured by the Hubble Space Telescope.

In principle, there is no limit to this technology. Different pieces of the same telescope could be flown on separate spacecraft. One space telescope following Earth could be mated through computer technology to an identical instrument on the opposite side of the Sun. Precisely aligned, this configuration would create an aperture as wide as the orbit of the Earth. Some astronomers want to place such arrays along Jupiter's orbital plane, providing an aperture of astonishing size.

With gaps in the glass, the resulting images will be incomplete. For certain types of astronomy, this provides an advantage. To capture an image of an Earth-like planet in a nearby solar system, astronomers need to nullify the light of the central star. The incomplete pattern of light produced by separate mirrors with gaps between them does this precisely. The image resembles an interference pattern; the instrument is called an interferometer and is the best tool for viewing extrasolar planets.

Scientists are also learning how to view electromagnetic radiation not previously observable from Earth. Light radiation comes in many forms–gamma rays, x-rays, ultraviolet, infrared, and radio waves, as well as the optical waves visible to human eyes. Much of this radiation fails to penetrate the Earth's atmosphere. Gamma rays and x-rays must be studied from space. In 1991, shuttle astronauts deployed the Compton Gamma Ray Observatory, and in 1999, the Chandra X-Ray Observatory.

The ability of scientists to view the full electromagnetic spectrum, as well as their ability to construct telescopes with divided mirrors, is revolutionizing human understanding of the cosmos. Humans are peering into sectors of the universe that no one would otherwise be able to see. They are creating a perspective once thought to belong only to the gods.

Space telescopes. Astronauts photographed the Compton Gamma Ray Observatory as they deployed it from the space shuttle Atlantis in 1991 (top). The Chandra X-Ray Observatory, shown here in an artist's depiction, was launched in 1999 (bottom). These first-generation observatories are being replaced by space telescopes of ever-increasing power.

Artist's depiction of the evolution of the universe, 1989.
In the painting "In the Beginning Nothing Became Everything," Paul Hudson portrays the unfolding of the universe from the big bang to the deployment of the Hubble Space Telescope. The universe has evolved to the point where at least one planet has produced creatures capable of observing it all.

Astonishing Views With these instruments, scientists can investigate how the universe began: they will view creation. The best image available through the Hubble Space Telescope captures light that left primordial galaxies more than 10 billion years ago. One picture, produced in 1996, shows an image of galaxies that may have formed no more than 1 billion years after the universe began. The galaxies are so far away that the light from them has taken almost the age of the universe to reach Earth.

Scientists want to observe events that occurred even earlier. In 1989 NASA scientists launched the Cosmic Background Explorer satellite (COBE). This special space telescope produced an image of the universe when it was only 300,000 years old. COBE captured variations in the temperature of what scientists call background radiation, remnants of the moment when the universe emerged from a dense fog of opaque light and became transparent. The instrument revealed the seeds from which the first galaxies formed.

With a space-based interferometer, scientists may be able to pierce that primordial fog. Scientists believe that massive events, such as the collision of two black holes, create waves in gravitational fields that can be measured with delicate instruments. In theory, such ripples penetrate all intervening material. They would not be screened out by phenomena that otherwise block telescopic views. Gravitational waves from the largest event of all–the creation of the universe–may still permeate the cosmos. A properly designed space interferometer could gather this information, observing evidence from the moment of creation, just one second after it occurred.

Scientists also want to view the formation of primordial stars. The early universe consisted principally of hydrogen and helium, from which the first stars formed. Elements essential to life–carbon, oxygen, iron, and other heavy elements–did not yet exist. Scientists believe that such elements are forged in stellar furnaces through the fusion process that powers stars. Most stars die and collapse into white dwarfs, but a few massive giants explode and blow these precious elements into space. From such debris new solar systems form, and eventually any life on them. Using space telescopes that capture long wavelength radiation, astronomers hope to pierce the gas and dust veils that obscure stellar nurseries. Hubble's successor, the Next Generation Space Telescope, will observe primordial stars and galaxies. New x-ray and gamma ray telescopes will record their death throes. One NASA group characterizes the effort as "a search for the eggs from which our stellar geese were hatched."

Around the stars planets appear. Scientists believe that planets surround many stars, even ancient ones located in globular star clusters where Earth-like elements do not abound. In the twentieth century, astronomers located extrasolar planets by recording tiny variations in the position and brightness of their parent stars as the planets orbited them. These planets cannot be seen through conventional telescopes. They can be viewed, however, through space-based interferometers as soon as such instruments are deployed. Such planet finders will screen out the glare of central stars and capture images of extrasolar planets in ever increasing detail. Scientists will examine the processes by which such planets coalesce and evolve into possibly habitable spheres.

Using space-based instruments, scientists will probe the nature of black holes. A great deal of matter collapsing into a small area produces gravitation forces so strong that even light fails to escape its pull. In the last decade of the twentieth century, space telescopes began to capture images of the matter surrounding black holes. The rapid movement of this material produces a distinctive signature–jets of matter shooting out at right angles from the disk of material being dragged in.

The gravitational force of a black hole distorts matter, space, and time. Time on a clock thrown into a black hole slows down. Exotic objects like black holes provide an excellent laboratory for testing the laws of nature as humans conceive them. If the laws work under the extreme conditions created around black holes, they work anywhere. Astronomers want to launch squadrons of telescopes to study black holes through the whole electromagnetic spectrum–visible light, infrared, x-ray, gamma ray, and radio wave–as well has through the gravity waves such objects may produce.

The universe is full of strange objects. Quasars emit more energy than the entire Milky Way galaxy from a region not much larger than our solar system. They are not stars. Scientists believe that quasars may be fueled by supermassive black holes. Satellites designed to detect clandestine nuclear explosions on Earth by the gamma rays they emit record similar events in outer space. These are not the result of weapons dispatched from Earth, but gamma ray bursts from the universe beyond. The bursts, lasting only one second or so, originate from objects outside the Milky Way, billions of light-years away. The "gamma ray bursters," as they are called, produce more energy in a short period of time than any other object in the universe, except for the big bang.

Pulsars, or pulsating radio sources, emit beams of energy that sweep by Earth with the predictable regularity of a beacon from a lighthouse tower. The first scientists to discover them thought they might be signals from an intelligent civilization. The beams, in fact, are emissions from spinning neutron stars. At the center of the Milky Way galaxy is a strange glowing cloud. Recorded by NASA's Compton Gamma Ray Observatory, it appears to be the result of matter colliding with antimatter.

Scientists will launch better telescopes to study objects like black holes, quasars, gamma ray bursters, antimatter, gravitational waves, and cosmic rays. The phenomena scientists discover will lead to laws of physics as yet unknown.

"The Hubble images, what are they good for, pragmatically? Is there any reason for a Hubble Telescope? What is the spinoff? What do we care about cosmology? What do we care about those nebulae? What do we care about star death and planets? Why do we care? Astronomy is for the sake of astronomy only . . . That's why it's so nice. That's why it's so delicious."

—Astronaut Story Musgrave, 1998.

Peering into Infinity At some point in the first half of the twenty-first century, scientists at a ground control station in the United States will send commands to a squadron of space telescopes traveling a great distance from Earth. On command, the individual, free-flying telescopes will point toward the same spot in

Photograph of the early universe, 1995.
The galaxies captured in this Hubble Space Telescope image are seen as they existed 10 billion years ago. Some may have formed less than 1 billion years after the big bang.

Images of black holes, 1994–2000.
Black holes were only theoretical until space telescopes confirmed their existence. Two artistic depictions show the distinctive streams of electrons ejected from the swirls of matter being sucked into black holes (top left and right). An image from the Chandra X-Ray Observatory reveals a relatively cool blue dot in the center of the Andromeda galaxy (bottom left). The dot houses a supermassive black hole with the mass of 30 million suns. The "event horizon" for a black hole appears in a photograph taken through the Hubble Space Telescope of galaxy M87 (bottom right). Measurements of the rotating disk confirm that it must contain a massive black hole at its hub.

opposite
Artist's concept of an extrasolar planet and star, 1999.
Perhaps as many as 100 million of the Sun-like stars in our galaxy harbor close-orbiting gas giants like Jupiter. Some may approach so close that their stars gobble them up. The next generation of space telescopes will capture images of extrasolar planets, though hardly in this much detail.

the sky. They will hold that point for a long period of time, perhaps a week or more, recording light waves photon by photon until an entire image forms. The image will change the way humans view the universe.

Perhaps the telescopes will capture a detailed image of material falling into a black hole. Perhaps they will record a feature from the big bang. It is quite possible that the instruments will take the first picture of a blue-and-white planet very much like Earth in another solar system.

The more of the universe that scientists comprehend, the stranger it appears. No one can predict what scientists will find. To gain perspective, consider how much astronomers have learned in the last fifty years.

In 1950, just a few years before space exploration began, no human had ever seen the backside of the Moon. No person had observed the surface of Venus or the landscape of Mars. No spacecraft had recorded the eruption of volcanoes on Io, one of Jupiter's moons, or watched a large comet slam into another planet, as Shoemaker-Levy 9 did to Jupiter in 1994.

In 1950, scientists had not yet discovered quasars; they were first detected in 1960. Pulsars were not discovered until 1967. No one had confirmed the existence of a black hole.

Fifty years ago many scientists believed in the concept of a "steady-state" universe, a theory that purported to show how matter was being continuously formed. No one had detected cosmic background radiation, the residual glow that provided the best twentieth-century evidence for the big bang.

Fifty years of scientific investigation with instruments of increasing sensitivity produced this yield. Fifty more years will transform the science fiction

"In the future, we will continue our exploration of the solar system and beyond. However, this exploration will proceed in ways that would have surprised, and I think fascinated, Wernher von Braun. The von Braun paradigm—that humans were destined to physically explore the solar system—was bold, but his vision was highly constrained by the technology of the day. For von Braun, humans were the most powerful and flexible exploration tool that he could imagine. Today, we have within our grasp technologies that will fundamentally redefine the exploration paradigm. We have the ability to put our minds where our feet can never go. We will soon be able to take ourselves—in a virtual way—anywhere from the interior of a molecule to the planets circling a nearby star."

—John Gibbons, White House science advisor, 1995.

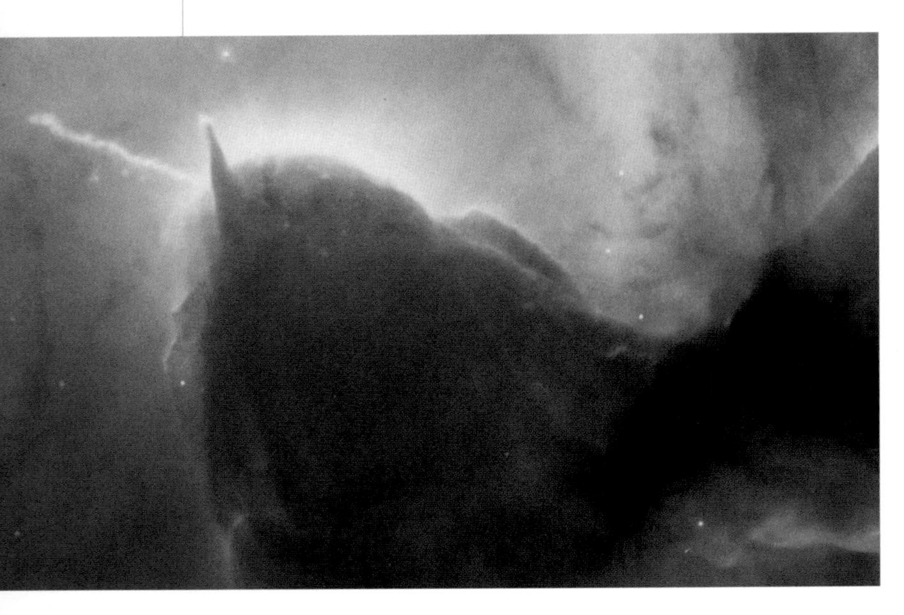

of today into the reality of tomorrow. Most of the material out of which the universe is formed is dark matter–it pulls on other objects but gives off no light. It is a material as yet unknown. Scientists do not know the shape of the whole universe. Perhaps it folds back on itself in ways that create an illusion of infinity. Wormholes may exist that provide short-cuts through space and time, even passageways to alternative universes. Scientists may discover where matter that disappears into black holes really goes.

Such investigation will expand scientific understanding. Around Earth, matter behaves within a narrow, well-defined range. No black holes rip human bodies apart; no gamma ray bursters strip the Earth's protective ozone layer away. This is just as well. Human life could not exist under the violent conditions that distinguish the more dangerous sectors of the universe.

Photographs of the universe, 1992–1999.
Telescopes have captured images that confirm the birth-place of stars (top), the manner in which exploding stars create the planetary nebulae from which new solar systems form (right), and the existence of quasars (bottom).

By studying those sectors, scientists increase the variety of conditions under which they view the laws of nature at work. Increased variety, which scientists call "variance," helps to establish causality. The best insights into the workings of nature come from breaking apart the very small, which can be done in laboratories on Earth, and understanding the very large. Looking deep into the cosmos, scientists perfect their understanding of how particles on Earth behave.

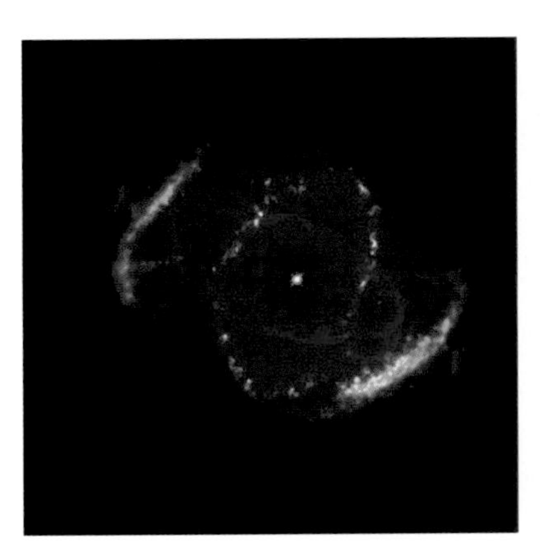

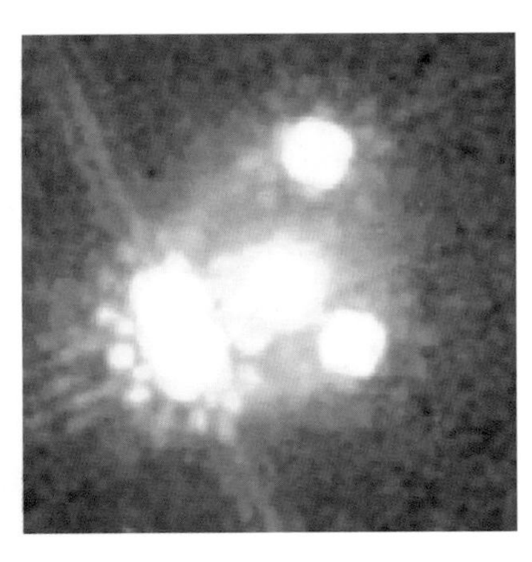

Studying exotic processes in space pushes the boundaries of technology. The body of knowledge required to fly new telescopes only begins with the creation of lightweight, computer-controlled mirrors, a substantial technical challenge by itself. Telescopes floating in the weightless environment of space are susceptible to image-smearing vibrations. The vibrations arise from the natural resonant frequencies of mechanisms within the instrument. Uncontrolled vibrations can destroy a large space telescope. Damping such vibrations is a design problem that can be solved only by developing sophisticated computer software programs that model operations in space.

On many missions, multiple space telescopes will fly in precise formations. This demands new techniques as well. Laser technology provides one avenue for maintaining the desired distance between free-flying instruments. Engineers are working to produce microthrusters that can adjust the position of flying telescopes by tiny degrees.

New views of the cosmos have cultural ramifications. To early humans, the night sky appeared like a dome over as much of Earth's surface as they could see. Since the stars moved across the sky, humans naturally assumed that Earth sat at the center of the creation. Some points of light traveled through the stars. The first telescopes revealed the true nature of these bodies, planets like Earth traveling around the Sun. The resulting spirit of inquiry spurred the Renaissance, the revival of art and learning that in turn made possible the scientific revolution. The products of science created spaceflight, from which humans gained their first view of Earth as seen from beyond. Norman Cousins once remarked that the greatest achievement of the race to the Moon "was not that man set foot on the Moon, but that he set eye on the Earth." Seeing the planet as a small blue-and-white globe spurred the environmental movement and the concept of humans as members of a global community. It, too, changed the way in which humans viewed the world.

Photographs of the universe, 1997–1998.

A strange cloud at the center of the Milky Way glows in gamma rays produced by the collision of matter and antimatter (below), while a pair of orbiting stars eject a bipolar planetary nebula (below left) ten times the diameter of our solar system.

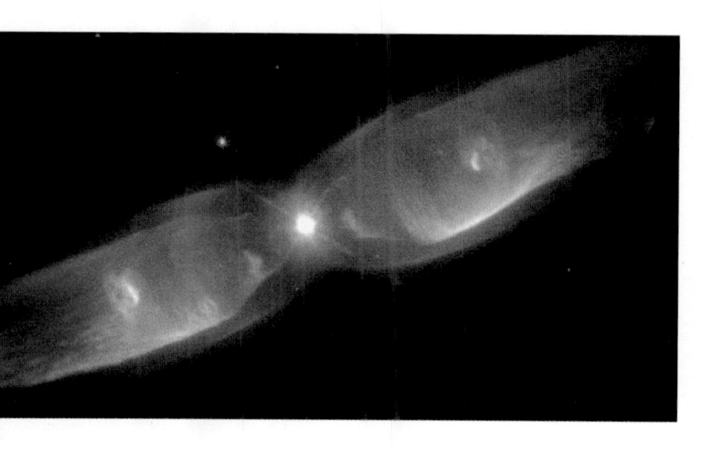

Soon this perspective will change again. Earth will shrink once more. Humans will come to view Earth as a small point in space, surrounded in three dimensions by a cosmic neighborhood of familiar stars and the planets around them. Robotic spacecraft, launched from Earth to explore the larger planets, have already ventured beyond the outer edge of the solar system. One even turned its camera back to capture a final montage. From beyond the orbits of the outermost planets, Earth appears as a pale blue dot nearly hidden in the glare of the Sun, a barely detectable home to creatures capable of building wandering machines.

Who knows what cosmic perspective the new point of view will patronize. Some think it will lead to a sense of insignificance as the world dwindles one more time. Not everyone agrees. Such a view is more likely to give humans a sense of the incredible significance of what they possess—a small and rocky sanctuary in a cosmos within which humans have been given the gift of understanding.

following pages

Artist's concept of the solar system, 1990.

Images from space are altering humanity's vision of Earth, reducing it to a rocky sanctuary hidden by the glare of its supporting sun. In 1983, Pioneer 10 became the first spacecraft to leave the solar system, an event depicted in this artist's rendering.

"If we do discover a complete theory [of the universe], it should in time be understandable in broad principle to everyone, not just a few scientists. Then we shall all, philosophers, scientists, and just ordinary people, be able to take part in the discussion of the question of why it is that we and the universe exist. If we find the answer to that, it would be the ultimate triumph of human reason—for then we would know the mind of God . . ."

—Stephen Hawking, physicist, 1988.

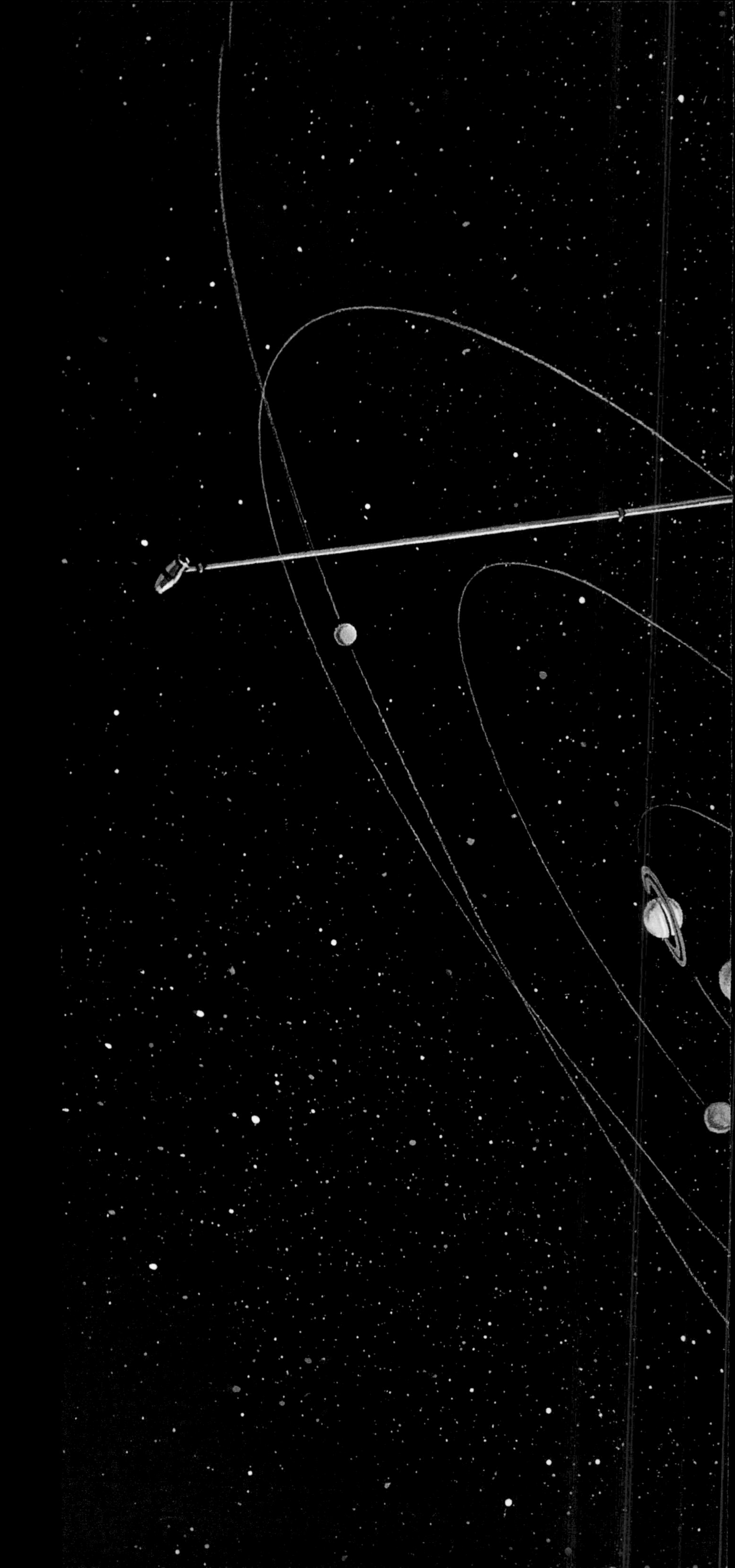

Epilogue: A Hopeful Future

During the first half of the twentieth century, rocket pioneers developed the means to hurl machines into outer space. Robert Goddard launched the first liquid-fuel rocket, a spindly affair that rose 41 feet into the air, from a Massachusetts farm in 1926. Twenty-three years later, Wernher von Braun and his rocket team fired a German V-2, topped with a WAC Corporal second stage, from White Sands, New Mexico, to a record altitude of 250 miles. The rocket reached outer space, then fell back to Earth.

By 1950, achievements in rocketry justified the bold claims of individuals such as these. Within a few years, they said, humans would fly into space and explore the solar system. They would build winged spaceships, construct large space stations, launch orbital observatories, organize expeditions to the Moon, travel to Mars, and discover extraterrestrial life. These were extraordinary conceptions to people living at that time. Many did not believe that they would occur.

To convince doubters of this fantastic future, space pioneers tied their vision to familiar images from the past. Space travel, they said, would rekindle the tradition of the American frontier. According to historian Frederick Jackson Turner, the western frontier in the United States closed in 1890. Rocket pioneers presented outer space as the "final frontier"–a new venue in which frontier values like inventiveness and cooperation could be renewed.

Rocket pioneers promised that space travel would revive excitement created by the era of terrestrial exploration. For the previous four hundred years, explorers had investigated seas and lands across the globe. The "golden age" of terrestrial discovery ended about 1926 with the first airplane flight over the South Pole. Humans had explored the Grand Canyon, discovered the source of the Nile, and hiked to the poles. Space travel promised to continue that tradition, with its commensurate discovery of new lands, exotic species, and revolutionary science.

Exploration enthusiasts presented space technology as a means by which the "winged gospel" would be extended higher still. Following the first flight by the Wright brothers in December 1903, aviation advocates made fantastic claims. In the future, they said, millions of people would fly through the air and be liberated by the experience. Rocket pioneers applied this familiar vision to the final frontier, promising a future of cheap and easy spaceflight available to ordinary people.

To a certain extent, much of the original vision came true. Within fifty years, humans constructed large rocketships, landed on the Moon, dispatched robots to search for life on Mars, built elaborate space observatories, and started work on an International Space Station.

In other respects, the vision bore little resemblance to actual events. Rocket scientists did build large, winged spaceships, but ordinary people could not fly on them. Engineers tried to build an orbital space station, but it proved more costly and difficult than the first pioneers proclaimed. Humans went to the Moon, but they failed to stay. Surveying Venus and Mars, scientists found conditions more hostile to the maintenance of life than anyone had dreamed.

An intrinsically new activity, space exploration held both disappointment and surprise. Few people looking back at the discovery of species on Earth anticipated how hard the search for extraterrestrial life would be. The absence of surface life on Mars was a terrible disappointment to people raised on Martian tales.

None of the visionaries anticipated the ease with which earthlings could communicate with spacecraft on long voyages. Visionaries recalled the isolation experienced by sea captains on long terrestrial journeys and assumed that spacecraft commanders would face similar constraints.

Few people from the past foresaw the commercial potential of space. In his classic March 1952 article on the orbiting space station, Wernher von Braun made no mention of microgravity research, the principle function of the International Space Station.

No one in 1950 foresaw the extent to which pictures of the whole Earth would reshape human thinking about the biosphere. Few people predicted the degree to which space could be used to monitor the health of the globe.

Twenty-first-century space advocates continue to dream about winged spaceships, rotating space stations, lunar bases, and colonies on Mars. Some of these visions will come true. To a large extent, however, the motivating visions rest on a foundation made of sand. They draw their force from nostalgic memories of a past that, to a great extent, never existed. Space exploration is not the twenty-first-century equivalent of the Oregon Trail, certainly not in the romantic manner that modern people remember that episode.

Rooted as it is in the past, the original vision of space exploration fails to anticipate much of the future. Like other youthful fantasies, the original vision must yield to more mature ideas. Space is a realm for the extension of everyday activities that humans carry out on the surface of Earth and in the air. Humans use Earth for government, commerce, and science. So too these activities enter space. The dividing line between Earth and outer space is dimming as humans learn how to operate there.

Humans in the twenty-first century face challenges from a world that grows more crowded, more polluted, and often more dangerous. It is hard to imagine these issues being resolved by civilizations trapped on the surface of the globe. Space provides the necessary elbow room for combating these trends.

Most humans who live during the twenty-first century will reside on the surface of Earth. They will not flee to Martian settlements or orbiting space colonies. Increasingly, humans will use space as a platform for gathering information necessary to regulate the biosphere. They will use space to monitor misbehavior on Earth, whether that misconduct takes the form of environmental pollution or preparations for war.

Space will provide a window on the universe from which fantastic new discoveries can be made. Humans may well discover extraterrestrial life. They may set their eyes on the image of an Earth-like planet around a nearby star. They may discover some fantastic material that can be made only in a gravity-free realm. Perhaps they may discover some heretofore unknown principle of physics. Perhaps they will capture an image of the creation.

None of the forecasts contained in this book are preordained to occur. All are possible; none are guaranteed. An accident similar to those that befell the shuttle Challenger and the French Concorde could dampen public enthusiasm for space exploration and retard what is technically possible. A startling discovery of extraordinary significance could accelerate space investment, propelling civilization into unexpected realms. Only one feature of space exploration is inevitable: surprises will occur. Space is full of achievements, disappointments, and surprises. By going into space, humans learn what they do not know.

The first fifty years of space exploration were motivated by fantastic images. Those images helped build widespread support for the endeavor and convince an otherwise skeptical public that such things could occur. Space will continue to supply its share of fantastic views. For the next fifty years, however, space is likely to provide something more than the achievement of fantasy. Properly conducted, space exploration can provide a hopeful future. It can offer an important part of the means by which humans learn to live on a small and precious world by learning to live a bit beyond it, too.

Acknowledgments

In any book, numerous debts are incurred over the course of production. This is especially true of a breathless futurist survey such as this, which of necessity relies almost exclusively on the primary research of others. We acknowledge the support and encouragement of a large number of people associated with aerospace studies and want to thank many individuals who materially contributed to the completion of this project. Of course, we would never have taken on this project were it not for the encouragement and ideas provided by Laura Lovett and the other fine people at Chronicle Books.

In addition, several individuals read all or part of this manuscript or otherwise offered suggestions which helped us immensely. Our thanks are extended to the staff of the NASA History Division: Stephen E. Garber, who offered valuable advice; Jane Odom, who helped track down materials and correct inconsistencies; M. Louise Alstork, who provided excellent editorial advice, and Nadine Andreassen, who offered invaluable assistance. In addition to these individuals, we wish to acknowledge the following people who aided us in a variety of ways: Judith Allton, Richard Berendzen, Mike Chambers, Tom D. Crouch, Frederick C. Durant III, Donald C. Elder, Richard Faust, Jack Frassanito, Thomas Fuller, Richard P. Hallion, Michael Hawes, T.A. Heppenheimer, Francis T. Hoban, Adam J. Hoffman, Nancy M. House, Dennis R. Jenkins, Sylvia K. Kraemer, W. Henry Lambright, John M. Logsdon, John L. Loos, Robert McCall, Pamela E. Mack, Ron Miller, John E. Naugle, Allan A. Needell, Regan Newport, Arnauld S. Nicogossian, Frederick I. Ordway III, David S.F. Portree, Pat Rawlings, Robert W. Smith, Rick W. Sturdevant, Bert Ulrich, and Joni Wilson. Our thanks also go to Jay Schaefer and Judith Dunham for their persistence in editing and seeing this manuscript through publication.

ROGER D. LAUNIUS
HOWARD E. MCCURDY
Washington, D.C.

A Library of Basic Books on Space

Barnes-Svarney, Patricia. *Asteroid: Earth Destroyer or New Frontier?* New York: Plenum Press, 1997. Discusses the challenge of defending the planet from asteroid impacts.

Bilstein, Roger E. *Flight in America: From the Wrights to the Astronauts.* Baltimore: Johns Hopkins University Press, 1984; paperback reprint, 1994. A superb synthesis of the origins of American air and space activities.

Boss, Alan. *Looking for Earths: The Race to Find New Solar Systems.* New York: John Wiley and Sons, 1998. An overview, from early theories of planetary formation to modern extrasolar planet searches.

Bromberg, Joan Lisa. *NASA and the Space Industry.* Baltimore: Johns Hopkins University Press, 1999. An important study of the government-industry relationship.

Burrows, William E. *This New Ocean: The Story of the First Space Age.* New York: Random House, 1998. A comprehensive history of forty years of spaceflight.

Caiden, Martin, and Jay Barbree, with Susan Wright. *Destination Mars: In Art, Myth, and Science.* New York: Penguin Studio, 1997. A beautifully illustrated overview of this fascinating planet.

Chaikin, Andrew. *A Man on the Moon.* New York: Viking, 1994. The story of America's voyages to the Moon, in paperback or in a richly illustrated, three-volume set.

Crosswell, Ken. *Planet Quest: The Epic Discovery of Alien Solar Systems.* New York: Free Press, 1997. A detailed analysis of the efforts to find extrasolar planets.

Crouch, Tom D. *Aiming for the Stars: The Dreamers and Doers of the Space Age.* Washington, D.C.: Smithsonian Institution Press, 1999. A history, from science fiction to twentieth-century rocketeers.

Dick, Steven J. *The Biological Universe: The Twentieth Century Extraterrestrial Life Debate and the Limits of Science.* New York: Cambridge University Press, 1996. An excellent analysis of the idea that life exists on other worlds.

Disch, Thomas M. *The Dreams Our Stuff is Made Of: How Science Fiction Conquered the World.* New York: The Free Press, 1998. A powerful analysis of the relationship between science fiction and science fact.

Drake, Frank, and Dava Sobel. *Is Anyone Out There? The Scientific Search for Extraterrestrial Intelligence.* New York: Delacorte Press, 1993. A superb discussion of the issues surrounding the search for extraterrestial intelligence.

Eckart, Peter, ed. *The Lunar Base Handbook: An Introduction to Lunar Base Design, Development, and Operations.* New York: McGraw Hill, 1999. How to develop a lunar base in the twenty-first century.

Fleming, James Rodger. *Historical Perspectives on Climate Change.* New York: Oxford University Press, 1998. A historical analysis of the global warming debate.

Goldsmith, Donald. *The Hunt for Life on Mars.* New York: E. P. Dutton, 1997. Written in the aftermath of the Mars meteorite announcement, this book explores the probabilities of life on Mars.

———. *The Runaway Universe: The Race to Discover the Future of the Cosmos.* Cambridge, Mass.: Perseus Books, 2000. An easy-to-understand discussion of cosmology and the science that has influenced its development.

Gouré, Daniel, and Christopher M. Szara, ed. *Air and Space Power in the New Millennium.* Washington, D.C.: Center for Strategic and International Studies, 1997. An important discussion of U.S. plans to base weapons in space.

Greeley, Ronald, and Raymond Batson. *The NASA Atlas of the Solar System.* New York: Cambridge University Press, 1996. The landmark book on space science at the end of the twentieth century.

Gribbin, John, and Simon Goodwin. *Origins: Our Place in Hubble's Universe.* Woodstock, New York: The Overlook Press, 1998. A "family photo album" of the solar system, with a collection of color images taken by telescopes and spacecraft.

Gribbin, John, and Mary Gribben. *Fire on Earth: In Search of the Doomsday Asteroid.* New York: Simon and Schuster, 1996. A popular discussion of the problems and possibilities of planetary protection.

Hall, R. Cargill, and Jacob Neufeld, ed. *The U.S. Air Force in Space: 1945 to the 21st Century.* Washington, D.C.: USAF History and Museums Program, 1998. A collection of short essays on the varied missions of the military in space.

Handberg, Roger. *Seeking New World Vistas: The Militarization of Space.* Westport, Conn.: Praeger Publishers, 2000. A study of the policy issues associated with arming the heavens.

Harland, David M. *The Space Shuttle: Roles, Missions and Accomplishments.* Chicester, England: Wiley-Praxis, 1998. The most sophisticated of any shuttle history to appear to date.

Harrison, Albert A. *After Contact: The Human Response to Extraterrestrial Life.* New York: Plenum Press, 1997. What would happen if humans made contact with extraterrestrial life.

Heppenheimer, T. A. *Countdown: The History of Space Exploration.* New York: John Wiley & Sons, 1997. A general history of spaceflight worldwide–quirky and entertaining.

Johnson, Dana J., Scott Pace, and C. Bryan Gabbard. *Space: Emerging Options for*

National Power. Santa Monica, Cal.: RAND, 1998. With the Cold War over and commercial use of space growing, the authors reevaluate military operations in space.

Koerner, David, and Simon LeVay. *Here There Be Dragons: The Scientific Quest for Extraterrestrial Life.* New York: Oxford University Press, 2000. A full assessment of astrobiology.

Kolb, Rocky. *Blind Watchers of the Sky: The People and Ideas that Shaped our View of the Universe.* Reading, Mass.: Addison-Wesley Publishing, 1996. An award-winning popular discussion of cosmology.

Launius, Roger D. *Frontiers of Space Exploration.* Westport, Conn.: Greenwood Press, 1998. A short history of space exploration along with key documents and biographical sketches.

———. *NASA: A History of the U.S. Civil Space Program.* Melbourne, Fla.: Krieger Publishing, 1994. NASA's role in the exploration of space.

Launius, Roger D., and Howard E. McCurdy, ed. *Spaceflight and the Myth of Presidential Leadership.* Urbana: University of Illinois Press, 1997. A collection of essays on space-age presidents and their role in shaping space policy.

Launius, Roger D., with Bertram Ulrich. *NASA & the Exploration of Space.* New York: Stewart, Tabori, and Chang, 1998. A history of NASA's role in space exploration heavily illustrated with works from the NASA art program.

Lewis, John S. *Mining the Sky: Untold Riches from the Asteroids, Comets, and Planets.* Reading, Mass.: Addison-Wesley Publishing, 1996. A discussion of how to carry out space-mining operations.

———. *Rain of Iron and Ice: The Very Real Threat of Comet and Asteroid Bombardment.* Reading, Mass.: Addison-Wesley Publishing, 1997. Past and future dangers of extraterrestrial impacts.

———. *Worlds Without End: The Exploration of Planets Known and Unknown.* Reading, Mass.: Perseus Books, 1998. An excellent discussion of the science of planet hunting and exploration.

Lewis, John S., and Ruth A. Lewis. *Space Resources: Breaking the Bonds of Earth.* New York: Columbia University Press, 1987. An account of the possibilities for commercial enterprise in space.

Ley, Willy. *Rockets, Missiles, and Men in Space.* New York: Viking Press, 1968. The fourth and final edition of the work first published as *Rockets,* one of the most significant textbooks available in the midtwentieth century on the possibilities of space travel.

Ley, Willy, and Chesley Bonestell. *The Conquest of Space.* New York: Viking, 1949. Illustrated with Bonestell's striking moonscapes and impressive spacecraft, this book suggests that spaceflight is both feasible and desirable as a way of accomplishing the human destiny.

Light, Michael. *Full Moon.* New York: Alfred A. Knopf, 1999. More than 120 stunning images from Apollo expeditions, supported by a narrative of breathtaking immediacy and authenticity.

Logsdon, John. *The Decision to Go to the Moon.* Cambridge, Mass.: MIT Press, 1970. An elaborate history of the events that led to President Kennedy's monumental decision.

Logsdon, John M., ed. *Exploring the Unknown: Selected Documents in the History of the U.S. Civil Space Program.* 4 Vols. Washington, D.C.: NASA Special Publication-4407, 1995–1999. An essential reference work containing more than 500 key documents affecting space policy in the twentieth century.

Mather, John, and John Boslough. *The Very First Light: The True Inside Story of the Scientific Journey Back to the Dawn of the Universe.* New York: Basic Books, 1996. A solid account of NASA's Cosmic Background Explorer, written by the project's chief scientist.

McCurdy, Howard E. *Inside NASA: High Technology and Organizational Change in the U.S. Space Program.* Baltimore: Johns Hopkins University Press, 1993. The evolution of NASA's culture from the creation of the agency to the 1990s.

———. *Space and the American Imagination.* Washington, D.C.: Smithsonian Institution Press, 1997. A significant analysis of the relationship between popular culture and public policy.

———. *The Space Station Decision: Incremental Politics and Technological Choice.* Baltimore: Johns Hopkins University Press, 1990. A study of the political process that led to the decision in 1984 to build the International Space Station.

McDougall, Walter A. ...*The Heavens and the Earth: A Political History of the Space Age.* New York: Basic Books, 1985; reprint edition, Baltimore: Johns Hopkins University Press, 1997. This Pulitzer Prize–winning book analyzes the space race to the Moon in the 1960s.

McLucas, John L. *Space Commerce.* Cambridge, Mass.: Harvard University Press, 1991. A useful analysis of the rise of business opportunities in space.

Morrison, David. *Exploring Planetary Worlds.* New York: Scientific American Books, 1993. A fine discussion of the planets and other bodies of the solar system using data returned from space.

Murray, Bruce C. *Journey into Space: The First Three Decades of Space Exploration.* New York: W. W. Norton, 1989. An excellent discussion of planetary science, written by the former director of the Jet Propulsion Laboratory.

Neal, Valerie, ed. *Where Next, Columbus? The Future of Space Exploration.* New York: Oxford University Press, 1994. Essays on space exploration as discovery, including future directions beyond the solar system.

O'Neill, Gerard K. *The High Frontier: Human Colonies in Space.* New York: Bantam Books, 1978. The classic statement on why humanity must create colonies in space and how to accomplish it.

Ordway, Frederick I., III, and Randy Lieberman, ed. *Blueprint for Space: From Science Fiction to Science Fact.* Washington, D.C.: Smithsonian Institution Press, 1992. A fine collection of essays and artwork linking popular culture to spaceflight.

Paine, Thomas O., et al. *Pioneering the Space Frontier: The Report of the National Commission on Space.* New York: Bantam Books, 1986. An official vision of the future of spaceflight from a presidential commission chaired by a former NASA administrator.

Penley, Constance. *NASA/Trek: Popular Science and Sex in America.* New York: Verso, 1997. A provocative analysis of NASA's treatment of women and the sexual subtext in *Star Trek.*

Preston, Richard. *First Light: The Search for the Edge of the Universe.* New York: Random House, 1996. A discussion of the scientific efforts to understand the origins of the universe.

Sagan, Carl. *Cosmos.* New York: Random House, 1980. A well-illustrated book written to accompany the stunning PBS series. Still an outstanding starting point for any research into the origin and evolution of the universe.

———. *Pale Blue Dot: A Vision of the Human Future in Space.* New York: Random House, 1994. The most sophisticated articulation of the space exploration imperative to appear since Wernher von Braun's work in the 1950s.

Schrunk, David G., Burton L. Sharpe, Bonnie L. Cooper, and Madhu Thangavelu. *The Moon: Resources, Future Development and Colonization.* New York: John Wiley and Sons, 1999. A compelling case for the establishment of a human base on the Moon.

Shapiro, Robert. *Planetary Dreams: The Quest to Discover Life Beyond Earth.* New York: John Wiley and Sons, 1999. Are we alone? A useful discussion of how life began on Earth and might be discovered elsewhere.

Sheehan, William. *The Planet Mars: A History of Observation & Discovery.* Tucson: University of Arizona Press, 1996. An excellent survey of how humans have acquired knowledge about the red planet, including succinct narratives of the Mariner and Viking missions.

Spires, David N. *Beyond Horizons: A Half Century of Air Force Space Leadership.* Peterson Air Force Base, Colo.: Air Force Space Command, 1997. An important history of military space operations.

Stafford, Thomas P., et al. *America at the Threshold: A Report of the Synthesis Group on America's Space Exploration Initiative.* Washington, D.C.: U.S. Government Printing Office, 1991. One attempt to plan a human expedition to Mars.

Stares, Paul B. *The Militarization of Space: U.S. Policy, 1945–1984.* Ithaca, New York: Cornell University Press, 1985. This seminal book suggests that national security needs drove much of the U.S. policy toward space exploration.

Stern, S. Alan, ed. *Our Worlds: The Magnetism and Thrill of Planetary Exploration as Described by Leading Planetary Scientists.* New York: Cambridge University Press, 1998. An insider's discussion of the major developments in planetary science.

Stine, G. Harry. *Halfway to Anywhere: Achieving America's Destiny in Space.* New York: M. Evans and Co., 1996. A sustained tract on the need to develop new launch vehicles.

———. *Living In Space: A Handbook for Work & Exploration Stations Beyond the Earth's Atmosphere.* New York: M. Evans and Co., 1997. A guide to the technologies necessary for humans to travel and work in space.

Stoker, Carol A., and Carter Emmart, ed. *Strategies for Mars: A Guide to Human Exploration.* San Diego: Univelt, Inc., 1996. An up-to-date work in a series on the exploration of Mars, detailing the technical, biological, and political challenges involved.

Turco, Richard. *Earth Under Siege: Air Pollution and Global Change.* New York: Oxford University Press, 1994. A frightening perspective on the environmental challenges ahead.

Tyson, Neil de Grasse. *Universe Down to Earth.* New York: Columbia University Press, 1995. An up-to-date explanation of the processes that govern the universe.

Von Benke, Matthew J. *The Politics of Space: A History of U.S.-Soviet/Russian Competition and Cooperation in Space.* Boulder, Colo.: Westview Press, 1997. An important analysis of the relationship between the two superpowers.

Walter, Malcolm. *The Search for Life on Mars.* Cambridge, Mass.: Perseus Books, 1999. Past and future efforts to find life on Mars.

Ward, Peter Douglas, and Donald Brownlee. *Rare Earth: Why Complex Life Is Uncommon in the Universe.* New York: Copernicus Books, 2000. Paleontologist Ward and astronomer Brownlee's rare Earth hypothesis predicts that simple, microbial life is widespread in the universe, while complex animal or plant life is extremely rare.

White, Frank. *The Overview Effect: Space Exploration and Human Evolution.* Reston,

Va.: American Institute of Aeronautics and Astronautics, 1998. Why humans must explore space.

Wilford, John Noble. *Mars Beckons: The Mysteries, the Challenges, the Expectations of Our Next Great Adventure in Space.* New York: Alfred A. Knopf, 1990. A superior explanation of the human fascination with Mars, including a discussion of early plans to send humans to the red planet.

Winter, Frank H. *Rockets into Space.* Cambridge, Mass.: Harvard University Press, 1990. Perhaps the most useful synthesis on rocketry available in English, written by the rocket curator at the National Air and Space Museum.

Zubrin, Robert. *Entering Space: Creating a Spacefaring Civilization.* New York: Jeremy P. Tarcher/Putnam, 1999. Taking a wide view of humanity's future in space, the author provides a road map for how humanity can expand its presence beyond Earth to the solar system and eventually interstellar space.

Zubrin, Robert, and Richard Wagner. *The Case for Mars: The Plan to Settle the Red Planet and Why.* New York: The Free Press, 1996. The most detailed explanation available of how to get humans to Mars.

Credits

Artwork

Aerial Images, Inc.: 123.

Agricultural Research Service, USDA: 113.

Bonestell Space Art: 2–3; 16–17; 18; 20; 24; 25; 27 top, center, and bottom; 28 top; 30 top left and top right; 30–31 bottom; 54; 56–57; 104; 132 left; 134; 152.

Bristol Spaceplanes, Ltd.: 114.

Robert McCall: 118; 128.

NASA: 1; 4–5; 6–7 (artwork by Pat Rawlings); 8–9 (artwork by Bryn Barnard); 10–11; 12; 21; 22; 26; 28 bottom; 31 top left (artwork by Wilson Hurley); 31 top right; 34; 37; 40–41; 43 (artwork by Pat Rawlings); 44 (artwork by Pat Rawlings); 46; 47; 49 top and bottom; 50 (artwork by Pat Rawlings); 51 (artwork by Mark Dowman and Doug McLeod); 53 (artwork by Greg Bacon); 55; 58, 59; 60; 61; 62 (artwork by Pat Rawlings); 64–65 (artwork by Jack Frassanito); 66–67 (artwork by Jack Frassanito); 68–69 (artwork by Ren Wicks); 70; 71; 73; 74 left; 74 right (artwork by Rolf Klep); 79; 80; 82 (artwork by Pat Rawlings); 83; 84; 85; 86; 87 (artwork by Jack Frassanito); 89; 90; 92; 93 (artwork by Leslie Bossinas); 95; 96; 97; 98; 101 top; 101 bottom (artwork by Pat Rawlings); 102; 103 left; 103 right (artwork by Pat Rawlings); 105 (artwork by Davidson); 108; 110; 112; 115; 117 (artwork by Pat Rawlings); 130; 132–33; 135; 138; 139; 140–41; 142 left and right; 145; 147; 148; 149; 151; 153; 155; 156–57 (artwork by Paul Hudson); 158; 160 (top left artwork by James Gitlin); 161 (artwork by James Gitlin); 162; 163; 164–65 (artwork by Lou Nolan).

National Air and Space Museum, Smithsonian Institution: 36.

National Archives and Records Administration: 124; 126; 127 top.

Courtesy of RIA Novosti, 120.

© Roger Ressmeyer/Corbis: 76.

Spacehab, Inc.: 106.

TRW, Inc.: 121; 127 bottom.

Sidebars

Grateful acknowledgment is made to the following for permission to reprint previously published material.

pages 8 and 50: "The Quest for Intelligent Life in Space is Just Beginning," by Carl Sagan. ©1978 by Carl Sagan. Reprinted with permission from the Estate of Carl Sagan and with permission from *Smithsonian* magazine.

page 102: "Robots vs. Humans: Who Should Explore Space," by Paul D. Spudis. ©1999 *Scientific American.* Reprinted with permission from Paul D. Spudis and with permission from *Scientific American* magazine.

page 111: "Grand Challenges for Space Exploration," by Wesley T. Huntress, Jr. ©1999 *Space Times: Magazine of the American Astronautical Society.* Reprinted with permission from Wesley T. Huntress, Jr., and with permission from *Space Times* magazine.

page 138. "Single Room, Earth View," by Sally Ride. ©1986 *Air & Space/Smithsonian.* Reprinted with permission from Sally Ride and with permission from *Air & Space/Smithsonian* magazine.

page 164: *A Brief History of Time,* by Stephen Hawking. ©1988 Bantam Books. Reprinted with permission from Stephen Hawking and with permission from Bantam Books, New York.

Index

Page numbers of illustrations are in italics.